For Reference

Not to be taken from this room

WITHDRAWN FROM
SOUTHERN OKLAHOMA
LIBRARY SYSTEM

A DICTIONARY OF PETROLEUM TERMS
THIRD EDITION

Edited by Jodie Leecraft

Published by
PETROLEUM EXTENSION SERVICE
Division of Continuing Education
The University of Texas at Austin
Austin, Texas
1983

Designed and illustrated by T. Rey Calderón

© 1983 by The University of Texas at Austin
All rights reserved
First Edition published 1976. Second Edition published 1979
Third Edition 1983
Printed in the United States of America
ISBN 0-88698-000-3
ISBN 0-88698-001-1 pbk.

PREFACE

This dictionary has resulted from many years of collecting and defining terms used in the petroleum industry. The project was initiated by John Woodruff when he assumed leadership of Petroleum Extension Service in 1944; after his death, it was continued by Bill Boyd, the new head. Under Boyd's direction, Ron Baker took over the tasks of updating and expanding the list of terms for the first edition of the dictionary, published in 1976. The second edition, with new terms added, was published in 1979. The present edition has been made even more extensive and, it is hoped, more useful for a larger number of people.

Terms included in the dictionary have come from many sources – writers and editors researching for new PETEX publications, individuals within the industry, instructors and coordinators of PETEX schools, and organizations with their own specific jargon. Special thanks are due the Gas Processors Association for the use of many terms from their GPA Glossary. Abbreviations and information regarding the SI system of units form additional matter that should be helpful.

Dictionaries such as this one must be updated and revised continually to stay abreast of the expanding, changing industry for and about which they are intended. We at PETEX hope that we will receive from our readers contributions of new terms and corrections or additions to the present ones. In that way we shall be able to publish new impressions and editions that will adequately serve the petroleum industry and those who are interested in it.

AAPG *abbr:* American Association of Petroleum Geologists.

AAPL *abbr:* American Association of Petroleum Landmen.

abaft *adv:* 1. toward the stern of a ship or mobile offshore drilling rig. 2. behind. 3. farther aft than. See *aft*.

abandon *v:* to cease producing oil and gas from a well when it becomes unprofitable. A wildcat well may be abandoned after it has proven nonproductive. Several steps are involved in abandoning a well: part of the casing may be removed and salvaged; one or more cement plugs are placed in the borehole to prevent migration of fluids between the different formations penetrated by the borehole; and the well is abandoned. In many states, it is necessary to secure permission from official agencies before a well may be abandoned.

abd, abdn *abbr:* abandoned; used in drilling reports.

abnormal pressure *n:* pressure exceeding or falling below the normal pressure to be expected at a given depth. Normal pressure increases approximately 0.465 psi per foot of depth (10.5 kPa per metre of depth). Thus, normal pressure at 10,000 feet is 4,650 psi; abnormal pressure at this depth would be higher or lower than 4,650 psi. See *pressure gradient*.

aboard *adv:* on or in a ship, offshore drilling rig, or helicopter.

abrasion *n:* wearing away by friction.

ABS *abbr:* American Bureau of Shipping.

absolute humidity *n:* the amount of moisture present in the air, usually expressed in grains of water per 100 cubic feet of air (milligrams of water per cubic metre of air). Compare *relative humidity*.

absolute permeability *n:* a measure of the ability of a single fluid (such as water, gas, or oil) to flow through a rock formation when the formation is totally filled (saturated) with the single fluid. The permeability measure of a rock filled with a single fluid is different from the permeability measure of the same rock filled with two or more fluids. Compare *effective permeability*.

absolute porosity *n:* the percentage of the total bulk volume of a rock sample that is composed of pore spaces or voids. See *porosity*.

Accumulator

absolute pressure *n:* total pressure measured from an absolute vacuum. It equals the sum of the gauge pressure and the atmospheric pressure corresponding to the barometer (expressed in pounds per square inch).

absolute temperature scale *n:* a scale of temperature measurement in which zero degrees is absolute zero. On the Rankine absolute temperature scale, in which degrees correspond to degrees Fahrenheit, water freezes at 492 degrees and boils at 672 degrees. On the Kelvin absolute temperature scale, in which degrees correspond to degrees Celsius, water freezes at 273 degrees and boils at 373 degrees. See *absolute zero*.

absolute zero *n:* a hypothetical temperature at which there is a total absence of heat. Since heat is a result of energy caused by molecular motion, there is no motion of molecules with respect to each other at absolute zero.

absorb *v:* 1. to take in and make part of an existing whole. 2. to recover liquid hydrocarbons from natural or refinery gas in a gas-absorption plant. The wet gas enters the absorber at the bottom and rises to the top, encountering a stream of absorption oil (a light oil) traveling downward over bubble-cap trays, valve trays, or sieve trays. The light oil removes, or absorbs, the heavier liquid hydrocarbons from the wet gas. See *bubble-cap trays, valve trays, sieve trays*.

absorbent *n:* also called absorption oil. See *absorption oil*.

absorber *n:* a vertical, cylindrical vessel that recovers heavier hydrocarbons from a mixture of predominantly lighter hydrocarbons. Also called absorption tower. See *absorb*.

absorber capacity *n:* the maximum volume of natural gas that can be processed through an absorber at a specified absorption oil rate, temperature, and pressure without exceeding pressure drop or any other operating limitation.

absorption *n:* the process of sucking up; taking in and making part of an existing whole. Compare *adsorption*.

absorption gasoline *n:* the gasoline extracted from natural gas by putting the gas into contact with oil in a vessel and subsequently distilling the gasoline from the heavier oil.

absorption oil *n:* a hydrocarbon liquid used to absorb and recover components from natural gas being processed.

absorption plant *n:* a plant that processes natural gas with absorption oil.

absorption-refrigeration cycle *n:* a mechanical refrigeration system in which the refrigerant is absorbed by a suitable liquid or solid. The most commonly used refrigerant is ammonia; the most commonly used absorbing medium is water. Compare *compression-refrigeration cycle*.

absorption tower *n:* also called absorber. See *absorber*.

abyssal *adj:* of or relating to the bottom waters of the ocean.

AC *abbr:* alternating current.

accelerator *n:* a chemical additive that reduces the setting time of cement. See *cement* and *cementing materials*.

accumulate *v:* to amass or collect. When oil and gas migrate into porous formations, the quantity collected is called an accumulation.

accumulator *n:* 1. a vessel or tank that receives and temporarily stores a liquid used in a continuous process in a gas plant. See *drip accumulator*. 2. on a drilling rig, the storage device for nitrogen-pressurized hydraulic fluid, which is used in closing the blowout preventers. See *blowout preventer control unit*.

acetic acid *n:* an organic acid compound sometimes used to acidize oilwells. It is not as corrosive as other acids used in well treatments. Its formula is $C_2H_4O_2$, or CH_3COOH.

acetylene welding *n:* a method of joining steel components in which acetylene gas and oxygen are mixed in a torch to attain the high temperatures necessary for welding.

acid *n:* any chemical compound, one element of which is hydrogen, that dissociates in solution to produce free hydrogen ions. For example, hydrochloric acid, HCl, dissociates in water to produce hydrogen ions, H^+, and chloride ions, Cl^-. This reaction is expressed chemically as $HCl \rightleftharpoons H^+ + Cl^-$. See *ion*.

acid brittleness *n:* low ductility of a metal due to its absorption of hydrogen gas. Also called hydrogen embrittlement.

acid fracture *v:* to part or open fractures in productive, hard limestone formations by using a combination of oil and acid or water and acid under high pressure. See *formation fracturing*.

acid gas *n:* a gas that forms an acid when mixed with water. In petroleum production and processing, the most common acid gases are hydrogen sulfide and carbon dioxide. They both cause corrosion, and hydrogen sulfide is very poisonous. See *sour gas, sour corrosion, sweet corrosion.*

acidity *n:* the quality of being acid. Relative acid strength of a liquid is measured by pH. A liquid with a pH below 7 is acid. See *pH value.*

acidize *v:* to treat oil-bearing limestone or other formations with acid for the purpose of increasing production. Hydrochloric or other acid is injected into the formation under pressure. The acid etches the rock, enlarging the pore spaces and passages through which the reservoir fluids flow. The acid is held under pressure for a period of time and then pumped out, after which the well is swabbed and put back into production. Chemical inhibitors combined with the acid prevent corrosion of the pipe.

acid stimulation *n:* a well stimulation method using acid. See *acidize.*

acid treatment *n:* a method by which petroleum-bearing limestone or other formations are put into contact with an acid to enlarge the pore spaces and passages through which the reservoir fluids flow.

acoustic log *n:* a record of the measurement of porosity done by comparing depth to the time it takes for a sonic impulse to travel through a given length of formation. The rate of travel of the sound wave through a rock depends on the composition of the formation and the fluids it contains. Because the type of formation can be ascertained by other logs, and because sonic transit time varies with relative amounts of rock and fluid, porosity can usually be determined in this way.

acoustic position reference *n:* a system consisting of a beacon positioned on the seafloor to transmit an acoustic signal, a set of three or four hydrophones mounted on the hull of a floating offshore drilling vessel to receive the signal, and a position display unit to track the relative positions of the rig and the drill site. Monitoring of the display unit aids in accurate positioning of the rig over the site.

acoustic signatures *n pl:* the characteristic patterns for various degrees of cement bonding between the casing and the borehole that appear on an oscilloscope display when a sonic cement bond log is made.

acoustic survey *n:* a well-logging method in which sound impulses are generated and transmitted into the formations opposite the wellbore. The time it takes for the sound impulses to travel through the rock is measured and recorded. Subsequent interpretation of the record (log) permits an estimation of the rock's porosity and fluid content to be made. The process is also called sonic logging. See *sonic logging.*

acoustic well logging *n:* the process of recording the acoustic characteristics of subsurface formations, based on the time required for a sound wave to travel a specific distance through rock. The rate of travel of the sound wave depends on the composition of the formation, its porosity, and its fluid content.

acre-foot *n:* a unit of volume often used in oil reservoir analysis, equivalent to the volume (as of oil or water) necessary to cover 1 acre to a depth of 1 foot.

acre-ft *abbr:* acre-foot.

across *prep:* over. The term usually relates conditions of fluid flow on one side of a piece of equipment to conditions on the opposite side (e.g., a pressure drop across a separator).

ACT *abbr:* automatic custody transfer.

activated charcoal *n:* a form of carbon characterized by a high absorptive and adsorptive capacity for gases, vapors, and colloidal solids.

activation *n:* a reaction in which an element has been changed into an unstable isotope during bombardment by neutrons.

adapter spool *n:* a joint to connect blowout preventers of different sizes or pressure ratings to the casinghead.

additive *n:* 1. in general, a substance added in small amounts to a larger amount of another substance to change some characteristic of the latter. In the oil industry, additives are used in lubricating oil, fuel, drilling mud, and casing cement. 2. in cementing, a substance added to cement to change the cement characteristics to satisfy specific conditions in the well. A cement additive may work as an accelerator, retarder, dispersant, or other reactant.

adhere *v:* to stick or bond to (as paint may adhere to a metallic surface).

adiabatic change *n:* a change in the volume, pressure, or temperature of a gas, occurring without a gain or loss of heat.

adiabatic expansion *n:* the expansion of a gas, vapor, or liquid stream from a higher pressure to a lower pressure with no change in enthalpy of the stream.

adjustable choke *n:* a choke in which the position of a conical needle or sleeve may be changed with respect to its seat to vary the rate of flow; may be manual or automatic. See *choke*.

adsorbent *n:* a solid used to remove components from natural gas being processed.

adsorption *n:* the adhesion of a thin film of gas or liquid to the surface of a solid. Liquid hydrocarbons are recovered from natural gas by passing the gas through activated charcoal, silica gel, or other solids, which extract the heavier hydrocarbons. Steam treatment of the solid removes the adsorbed hydrocarbons, which are then collected and recondensed. The adsorption process is also used to remove water vapor from air or natural gas. Compare *absorption*.

adsorption plant *n:* a plant that processes natural gas with an adsorbent.

aeration *n:* the injection of air or gas into a liquid or a solid such as sand. For example, air is injected into drilling liquid to reduce the density of the liquid.

aerobic *adj:* requiring free atmospheric oxygen for normal activity.

aerobic bacteria *n pl:* bacteria that require free oxygen for their life processes. Aerobic bacteria can produce slime or scum that accumulates on metal surfaces, causing oxygen-concentration-cell corrosion.

A-frame *n:* 1. a derrick or crane shaped like the letter A and used to handle heavy loads. 2. an A-shaped openwork structure that is the stationary and supporting component of the mast of a jackknife rig and to which the mast is anchored when it is in an upright or drilling position. 3. the uppermost section of a standard derrick, shaped like the letter A and used as a support in lifting objects such as the crown block to the water table.

aft *adv:* toward or near the stern of a ship or offshore drilling rig.

AGA *abbr:* American Gas Association.

age *v:* to allow cement to mature, or reach a stage harder than that of immediate setting. The process is sometimes called *curing*. See *cure*.

agitator *n:* a motor-driven paddle or blade used to mix the liquids and solids in drilling mud.

A-h *abbr:* ampere-hour.

AIME *abbr:* American Institute of Mining, Metallurgical, and Petroleum Engineers.

air-actuated *adj:* powered by compressed air, as are the clutch and the brake system in drilling equipment.

air-cooled exchanger *n:* an atmospheric fin-tube exchanger that utilizes air for cooling. Ambient air contacts the external fins by fan-forced or natural draft. Also called air-fin unit or aerial cooler.

air-cut *adj:* having inadvertent mechanical incorporation of air into a liquid system.

air diving *n:* diving in which a diver uses a normal atmospheric mixture of oxygen and nitrogen as a breathing medium. It is limited to depths less than 190 feet (58 m) because of the dangers of nitrogen narcosis; however, dives with bottom times of 30 minutes or less may be conducted to a maximum of 220 feet (67 m).

air drilling *n:* a method of rotary drilling that uses compressed air as the circulation medium. The conventional method of removing cuttings from the wellbore is to use a flow of water or drilling mud. Compressed air removes the cuttings with equal or greater efficiency. The rate of penetration is usually increased considerably when air drilling is used. However, a principal problem in air drilling is the penetration of formations containing water, since the entry of water into the system reduces the ability of the air to remove the cuttings.

air gap *n:* the distance from the normal level of the sea surface to the bottom of the hull or base of an offshore drilling platform.

air gun *n:* a hand tool that is powered pneumatically.

air hoist *n:* a hoist operated by compressed air; a pneumatic hoist.

air injection *n:* 1. the injection of air into a reservoir in a pressure maintenance or an *in situ* combustion project. 2. the injection of fuel into the combustion chamber of a diesel engine by means of a jet of compressed air.

air tube clutch *n:* a clutch containing an inflatable tube that, when inflated, causes the clutch to engage the driven member. When the tube is deflated, disengagement occurs.

alarm *n:* a warning device triggered by the presence of abnormal conditions in a machine or system. For example, a low-water alarm automatically signals when the water level in a vessel falls below its preset minimum. Offshore, alarms are used to warn personnel of dangerous or unusual conditions, such as an approaching hurricane.

aliphatic hydrocarbons *n pl:* hydrocarbons that are characterized by having a straight chain of carbon atoms. Compare *aromatic hydrocarbons*.

aliphatic series *n:* a series of open-chained hydrocarbons. The two major classes of the series are the series with saturated bonds and the series with unsaturated bonds.

alkali *n:* a substance having marked basic (alkaline) properties, such as a hydroxide of an alkali metal. See *base*.

alkaline (caustic) flooding *n:* a method of enhanced oil recovery in which alkaline chemicals such as sodium hydroxide are injected during a waterflood or combined with polymer flooding. The chemicals react with the natural acid present in certain crude oils to form surfactants within the reservoir. The surfactants enable the water to move additional quantities of oil from the depleted reservoir. See *micellar-polymer flooding*.

alkalinity *n:* the combining power of a base, or alkali, as measured by the number of equivalents of an acid with which it reacts to form a salt. Measured by pH, alkalinity is possessed by any solution that has a pH greater than 7. See *pH value*.

alkylation *n:* a process for manufacturing components for 100-octane gasoline. An alkyl group is introduced into an organic compound either with or without a catalyst. Now usually used to mean alkylation of isobutane with propene, butenes, or hexenes in the presence of concentrated sulfuric acid or anhydrous hydrofluoric acid.

allocation *n:* the distribution of oil or gas produced from a well per unit of time. In a state using proration, this figure is established monthly by its conservation agency.

allowable *n:* the amount of oil or gas legally permitted to be produced from a well per unit of time. In a state using proration, this figure is established monthly by its conservation agency. See *proration*.

alloy *n:* a substance with metallic properties composed of two or more elements in solid solution. See *ferrous alloy* and *nonferrous alloy*.

alpha particle *n:* one of the extremely small particles of an atom that is ejected from a radioactive substance (such as radium or uranium) as it disintegrates. Alpha particles have a positive charge.

alternating current *n:* an electric current that reverses its direction of flow at regular intervals.

alternator *n:* an electric generator that produces alternating current.

aluminum bronze *n:* an alloy of copper and aluminum that may also include iron, manganese, nickel, or zinc.

A-mast *n:* an A-shaped arrangement of upright poles, usually steel, used for lifting heavy loads. See *A-frame*.

ambient temperature *n:* the temperature of the medium by which an object is surrounded.

American Association of Petroleum Geologists *n:* a leading national industry organization headquartered in Tulsa, Oklahoma. Its official publication is the *AAPG Journal*.

American Association of Petroleum Landmen *n:* an international trade organization, founded in 1955 and based in Fort Worth, Texas, for landmen and related professionals.

American Bureau of Shipping *n:* an organization that sets standards and specifications for ships and ship equipment manufactured in the U.S.A.

American Gas Association *n:* a national trade association of the petroleum industry whose members are U.S. and Canadian distributors of natural, manufactured, and mixed gases. AGA provides information on sales, finances, utilization, and all phases of gas transmission and distribution.

American Institute of Mining, Metallurgical, and Petroleum Engineers *n:* a New York City-based parent group of the Society of Petroleum Engineers (SPE). The SPE, headquartered in Dallas, Texas, publishes the *Journal of Petroleum Technology*.

American Petroleum Institute *n:* founded in 1920, this national oil trade organization is the leading standardizing organization for oil field drilling and producing equipment. It maintains departments of transportation, refining, and marketing in Washington, D.C., and a department of production in Dallas. *adj:* (slang) indicative of a job being properly or thoroughly done (as, "His work is strictly API").

American Society for Testing and Materials *n:* an organization, based in Philadelphia, which sets guidelines for the testing and use of equipment and materials.

American Society of Mechanical Engineers n: a New York City-based organization whose equipment standards are sometimes used by the oil industry. Its official publication is *Mechanical Engineering*.

American Society of Safety Engineers n: an organization that establishes safety practices for several industries. It maintains a national office in Chicago.

amine n: any of several compounds employed in treating natural gas. Amines are generally used in water solutions to remove hydrogen sulfide and carbon dioxide from gas or liquid streams.

ammeter n: an instrument for measuring electric current in amperes.

amortization n: 1. the return of a debt (principal and interest) in equal annual installments. 2. the return of invested principal in a sinking fund.

ampere n: the fundamental unit of electrical current. The symbol for ampere is A.

ampere-hour n: a unit of electricity equal to the amount produced in 1 hour by a flow of 1 ampere. See *ampere*.

anaerobic bacteria n pl: bacteria that do not require free oxygen to live or are not destroyed by its absence. Under certain conditions, anaerobic bacteria can cause scale to form in water-handling facilities in oil fields or hydrogen sulfide to be produced from sulfates.

anchor n: any device that secures or fastens equipment. In downhole equipment, the term often refers to the tail pipe. In offshore drilling, floating drilling vessels are often secured over drill sites by large metal anchors like those used on ships.

anchor buoy n: a floating marker used in a spread mooring system to position each anchor of a semisubmersible rig or drill ship. See *spread mooring system*.

anchor pattern n: the pattern of minute projections from a metal surface produced by sandblasting, shot blasting, or chemical etching to enhance the adhesiveness of surface coatings.

anchor washpipe spear n: a fishing tool attached to washover pipe by means of slips and released from the pipe once the fish is engaged by the tool. It provides a way to perform a washover and retrieve a stuck fish off the bottom in one round trip. See *fish* and *washover pipe*.

angle azimuth indicator n: also called riser angle indicator. See *riser angle indicator*.

angle control section n: also called a crossover. See *crossover*.

angle of deflection n: in directional drilling, the angle, expressed in degrees, at which a well is deflected from the vertical by a whipstock or other deflecting tool.

angle of deviation n: also called drift angle and angle of drift. See *deviation*.

angle of drift n: also called angle of deviation and drift angle. See *deviation*.

angle of wrap n: the distance that the brake band wraps around the brake flange. Drawworks have an angle of wrap of 270 degrees or more.

angle sub n: also called bent sub. See *bent sub*.

angular unconformity n: an unconformity in which formations on opposite sides are not parallel. See *unconformity*.

anhydrite n: the common name for anhydrous calcium sulfate, $CaSO_4$. Anhydrite formations are sometimes encountered during drilling. See *calcium sulfate, anhydrous*.

anhydrous adj: without water.

aniline point n: the lowest temperature at which the chemical aniline and a solvent (such as the oil in oil-base muds) will mix completely. Generally, the oil of oil-base muds should have an aniline point of at least 150°F (66°C) to obtain maximum service life from the rubber components in the mud system.

anion n: a negatively charged ion; the ion in an electrolyzed solution that migrates to the anode. See *ion*. Compare *cation*.

annular blowout preventer n: a large valve, usually installed above the ram preventers, that forms a seal in the annular space between the pipe and the wellbore or, if no pipe is present, on the wellbore itself. Compare *ram blowout preventer*.

annular space n: 1. the space surrounding a cylindrical object within a cylinder. 2. the space around a pipe in a wellbore, the outer wall of which may be the wall of either the borehole or the casing; sometimes termed the annulus.

annular velocity n: the rate at which mud is traveling in the annular space of a drilling well.

annulus n: also called annular space. See *annular space*.

anode n: 1. one of two electrodes in an electrolytic cell; represented as the negative terminal of the cell, it is the area from which electrons flow. In a primary cell it is the electrode that is wasted or eaten away. 2. in cathodic protection systems, an electrode to which a positive potential of electricity is applied, or a sacrificial anode, which provides protection to a structure by forming one electrode of an electric cell.

anomaly *n:* a deviation from the norm. In geology, the term indicates an abnormality such as a fault or dome in a sedimentary bed.

anoxia *n:* an undersupply of oxygen reaching the tissues of the body, possibly causing permanent damage or death. Also called hypoxia.

anticline *n:* an arched, inverted-trough configuration of folded and stratified rock layers. Compare *syncline*.

anticlinal trap *n:* a hydrocarbon trap in which petroleum accumulates in the top of an anticline. See *anticline*.

antifreeze *n:* a chemical compound that prevents the water in the cooling system of an engine from freezing. Glycols are often used as antifreeze agents.

antiknock compound *n:* a substance (specifically, tetraethyl lead) added to the fuel of an internal-combustion engine to prevent detonation of the fuel. Antiknock compounds effectively raise the octane rating of a fuel so that the fuel burns properly in the combustion changer of an engine. See *tetraethyl lead* and *octane rating*.

AOSC *abbr:* Association of Oilwell Servicing Contractors.

API *abbr:* American Petroleum Institute.

API cement class *n:* a classification system for oilwell cements, defined in *API Spec 10A*.

API gamma ray unit *n:* the standard unit of gamma ray measurement. Standardization of this unit results from the normalization of the detector-measurement systems of all primary service companies in the API test pits at the University of Houston. The API gamma ray unit is defined as $1/200$ of the difference in log deflection between two zones of different gamma ray intensity. The test pit is constructed so that the average midcontinent shale will record about 100 API gamma ray units.

API gravity *n:* the measure of the density or gravity of liquid petroleum products on the North American continent, derived from specific gravity in accordance with the following equation:

$$\text{API gravity} = \frac{141.5}{\text{specific gravity}} - 131.5$$

API gravity is expressed in degrees, a specific gravity of 1.0 being equivalent to 10° API.

API neutron unit *n:* the standard unit of measurement for neutron logs. Standardization of this unit results from the calibration of each logging tool model in the API neutron test pit at the University of Houston.

aquifer *n:* 1. a rock that contains water. 2. the part of a water-drive reservoir that contains the water.

Archie's equation *n:* the formula for evaluating the quantity of hydrocarbons in a formation. The form of the equation depends on its specific use. The basic equation is

$$S_w^2 = \frac{aR_w}{\Phi^m R_t}$$

arc weld *v:* to join metals by utilizing the arc created between the welding rod, which serves as an electrode, and a metal object. The arc is a discharge of electric current across an air gap. The high temperature generated by the arc melts both the electrode and the metal, which fuse.

arenaceous *adj:* pertaining to sand or sandy rocks (such as arenaceous shale).

argillaceous *adj:* pertaining to a formation that consists of clay or shale (such as argillaceous sand).

armature *n:* a part made of coils of wire placed around a metal core, in which electric current is induced in a generator, or in which input current interacts with a magnetic field to produce torque in a motor.

aromatic hydrocarbons *n pl:* hydrocarbons derived from or containing a benzene ring. Many have an odor. Cyclic, or single-ring, aromatic hydrocarbons are the benzene series (benzene, ethylbenzenes, and toluene). Polycyclic aromatic hydrocarbons include naphthalene and anthracene. Compare *aliphatic hydrocarbons*.

artificial lift *n:* any method used to raise oil to the surface through a well after reservoir pressure has declined to the point at which the well no longer produces by means of natural energy. Sucker rod pumps, gas lift, hydraulic pumps, and submersible electric pumps are the most common forms of artificial lift.

asbestos felt *n:* a wrapping material, consisting of asbestos saturated with asphalt, which is one element of pipeline coatings.

ash *n:* noncombustible residue from the gasification or burning of coal or a heavy hydrocarbon.

ASME *abbr:* American Society of Mechanical Engineers.

asphalt-base oil *n:* See *naphthene-base oil*.

asphalt enamel *n:* an asphalt-base enamel applied as a coating to pipe that is to be buried. The asphalt is combined with finely ground mica, clay, soapstone, or talc and applied while hot. Combined with a subsequent wrapping of asbestos felt, this coating protects the buried pipe from corrosion.

asphaltic material *n:* one of a group of solid, liquid, or semisolid materials that are predominantly mixtures of heavy hydrocarbons and their nonmetallic derivatives and are obtained either from natural bituminous deposits or from the residues of petroleum refining.

ASSE *abbr:* American Society of Safety Engineers.

associated gas *n:* natural gas that overlies and contacts crude oil in a reservoir. Where reservoir conditions are such that the production of associated gas does not substantially affect the recovery of crude oil in the reservoir, such gas may be reclassified as nonassociated gas by a regulatory agency.

Association of Oilwell Servicing Contractors *n:* an organization that sets some of the standards, principles, and policies of oilwell servicing contractors. It is based in Dallas, Texas.

astern *adv or adj:* 1. at or toward the stern of a ship or an offshore drilling rig; abaft. 2. behind the ship or rig.

ASTM *abbr:* American Society for Testing and Materials.

ASTM distillation *n:* any distillation made in accordance with ASTM procedure. Generally it refers to a distillation test to determine the initial boiling point, the temperature at which percentage fractionations of the sample are distilled, the final boiling point, and quantity of residue.

athwart *prep or adv:* from side to side of a ship or offshore drilling rig.

atm *abbr:* atmosphere.

atmosphere *n:* a unit of pressure equal to the atmospheric pressure at sea level, 14.7 pounds per square inch (101.325 kPa). One atmosphere is equal to 14.7 psi or 101.325 kPa.

atmospheres absolute *n pl:* total pressure at a depth underwater, expressed as multiples of normal atmospheric pressure.

atmospheric pressure *n:* the pressure exerted by the weight of the atmosphere. At sea level, the pressure is approximately 14.7 psi (101.325 kPa), often referred to as 1 atmosphere.

atmospheric pressure cure *n:* the aging of specimens for test purposes at normal atmospheric pressure for a designated period of time under specified conditions of temperature and humidity.

atom *n:* the smallest quantity of an element capable of either entering into a chemical combination or existing alone.

atomize *v:* to spray a liquid through a restricted opening, causing it to break into tiny droplets and mix thoroughly with the surrounding air.

attapulgite *n:* a fibrous clay mineral that is a viscosity-building substance, used principally in saltwater-base drilling muds.

automated plant *n:* a plant that contains instruments for the measurement and control of temperatures, pressures, flow rates, and properties of resulting products and thereby makes necessary corrections in the plant operating conditions so as to maintain specification products. Such a plant contains shutdown and other automatic devices for minimizing damage in the absence of operating personnel.

automatic choke *n:* an adjustable choke that is power-operated to control pressure or flow. See *adjustable choke*.

automatic control *n:* a device that regulates various factors (such as flow rate, pressure, or temperature) of a system without supervision or operation by personnel. See *instrumentation*.

automatic custody transfer *n:* a system for automatically measuring and sampling oil or products at points of receipt or delivery. See *lease automatic custody transfer*.

automatic driller *n:* a mechanism used to regulate the amount of weight on the bit without requiring attendance by personnel. Automatic drillers free the driller from the sometimes tedious task of manipulating the drawworks brake in order to maintain correct weight on the bit. Also called an automatic drilling control unit.

automatic drilling control unit *n:* also called automatic driller. See *automatic driller*.

automatic fillup shoe *n:* a device that is installed on the first joint of casing and that automatically regulates the amount of mud in the casing. The valve in this shoe keeps mud from entering the casing until mud pressure causes the valve to open, allowing mud to enter the casing.

automatic gauge *n:* an instrument installed on the outside of a tank to permit observation of the depth of the liquid inside.

automatic pumping station *n:* an automatically operated station installed on a pipeline to provide pressure when a fluid is being transported.

automatic shutdown *n:* a system in which certain instruments are used to control or maintain the operating conditions of a process. If conditions become abnormal, this system automatically stops the process and notifies the operator of the problem.

automatic slips *n:* a device, operated by air or hydraulic fluid, that fits into the opening in the rotary table when the drill stem must be suspended in the wellbore (as when a connection or trip is being made). Automatic slips, also called power slips, eliminate the need for roughnecks to set and take out slips manually. See *slips*.

automation *n:* automatic, self-regulating control of equipment, systems, or processes. See *instrumentation*.

auxiliaries *n pl:* equipment on a drilling or workover rig that is not directly essential to the basic process of making hole, like the equipment used to generate electricity for rig lighting or the equipment used to facilitate the mixing of drilling fluid. While the auxiliaries are not essential to the drilling process, their presence on the rig is often necessary if drilling is to progress at a satisfactory rate.

auxiliary brake *n:* a braking mechanism, supplemental to the mechanical brake, that permits the lowering of heavy hook loads safely at retarded rates without incurring appreciable brake maintenance. There are two types of auxiliary brakes – the hydrodynamic and the electrodynamic. In both types, work is converted into heat, which is dissipated through liquid cooling systems. See *hydrodynamic brake* and *electrodynamic brake*.

auxiliary equipment *n:* equipment subsidiary or supplementary to the main equipment used in an operation.

avg *abbr:* average.

axial compression *n:* pressure produced parallel with the cylinder axis when casing hits a deviation in the hole or a sticky spot and stops. The force pushing down on the pipe causes axial compression.

azeotrope *n:* a mixture of liquids that distills without a change in composition. The azeotropic mixture exhibits a constant maximum or minimum boiling point, which is either higher or lower than that of its components.

azimuth *n:* 1. in directional drilling, the direction of the face of the deviation tool with respect to magnetic north, as recorded by a deviation instrument. 2. an arc of the horizon measured between a fixed point (such as true north) and the vertical circle passing through the center of an object.

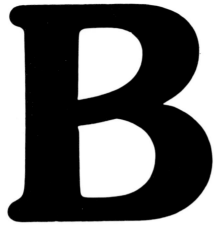

backout *v:* to overcome the positive electrical potentials of anodic areas in cathodic protection systems.

back-pressure *n:* 1. the pressure maintained on equipment or systems through which a fluid flows. 2. in reference to engines, a term used to describe the resistance to the flow of exhaust gas through the exhaust pipe.

back-pressure valve *n:* a valve used to regulate back-pressure on equipment or systems through which a fluid flows.

back up *v:* to hold one section of an object such as pipe while another section is being screwed into or out of it.

Blowout preventer

B *abbr:* bottom of; used in drilling reports.

babbitt *n:* metal alloy, either tin-based or lead-based, primarily used in friction bearings.

back-in unit *n:* a portable servicing or workover rig that is self-propelled, using the hoisting engines for motive power. Because the driver's cab is mounted on the end opposite the mast support, the unit must be backed up to the wellhead. See *carrier rig.*

back off *v:* to unscrew one threaded piece (such as a section of pipe) from another.

back-off joint *n:* a section of pipe with left-hand threads on one end and conventional right-hand threads on the other. In setting a liner, a back-off joint is attached to it so that the drill pipe may be disengaged from the liner by conventional right-hand rotation.

backup tongs *n:* the tongs used to back up the drill pipe as it is being made up into or taken out of the drill stem. See *make up, break out,* and *tongs.*

backup wrench *n:* any wrench used to hold a pipe or a bolt to prevent its turning while another length of pipe or a nut is being screwed into or out of it.

bacteria *n pl:* a large, widely distributed group of typically one-celled microorganisms. See *anaerobic bacteria* and *sulfate-reducing bacteria.*

bactericide *n:* anything that destroys bacteria.

baffle plate *n:* 1. a partial restriction, generally a plate, placed to change the direction, guide the flow, or promote mixing within a tank or vessel. 2. a device that is seated on the bit pin, in a tool joint, or in a drill pipe float, used to centralize the lower end of a go-devil while permitting the bypass of drilling fluid. The go-devil contains a surveying instrument.

bail *n:* a cylindrical steel bar (similar in form to the handle or bail of a bucket, but much larger) that supports the swivel and connects it to the hook. Sometimes, the two cylindrical bars that support the elevators and attach them to the hook are also called bails or links. *v:* to recover bottomhole fluids, samples, mud, sand, or drill cuttings by lowering a cylindrical vessel called a bailer to the bottom of a well, filling it, and retrieving it.

bailer *n:* a long cylindrical container, fitted with a valve at its lower end, used to remove water, sand, mud, drill cuttings, or oil from a well.

bailing drum *n:* the reel around which the bailing line is wound. See *bailing line*.

bailing line *n:* the cable attached to a bailer, passed over a sheave at the top of the derrick, and spooled on a reel.

ballast *n:* 1. for ships, water taken onboard into specific tanks to permit proper angle of repose of the vessel in the water, and to assure structural stability. 2. for mobile offshore drilling rigs, weight added to make the rig more seaworthy, increase its draft, or sink it to the seafloor. Seawater is usually used for ballast, but sometimes concrete or iron is used additionally to lower the rig's center of gravity permanently.

ballasted condition *n:* the condition of a floating offshore drilling rig when ballast has been added.

ballast movement *n:* a voyage or voyage leg made without any paying cargo in the vessel's tanks. To maintain proper stability, trim, or draft, seawater is usually carried during such movements. Also called ballast leg or ballast passage.

ballast tank *n:* any shipboard tank or tanker compartment normally used for carrying saltwater ballast. When these compartments or tanks are not connected with the cargo system, they are called segregated ballast tanks or systems.

ball bearing *n:* a bearing in which the journal turns upon loose hardened-steel balls that roll easily in a race and thus convert sliding friction into rolling friction. See *ball race*.

balled-up bit *n:* See *ball up*.

ball cock *n:* a device for regulating the level of fluid in a tank, consisting of a valve connected with a hollow floating ball, which by rising or falling shuts or opens the valve.

ball joint *n:* See *flex joint*.

ball race *n:* a track, channel, or groove in which ball bearings turn.

ball sealers *n:* balls made of nylon, hard rubber, or both and used to shut off perforations through which excessive fluid is being lost.

ball up *v:* 1. to collect a mass of sticky consolidated material, usually drill cuttings, on drill pipe, drill collars, bits, and so forth. A bit with such material attached to it is termed a balled-up bit. Balling up is frequently the result of inadequate pump pressure or insufficient drilling fluid. 2. in reference to an anchor, to fail to hold on a soft bottom, pulling out instead with a large ball of mud attached.

bar *v:* to move or turn (as a flywheel) by a bar used as a lever.

barefoot completion *n:* also called open-hole completion. See *open-hole completion*.

barge *n:* any one of many types of flat-decked, shallow-draft vessels, usually towed by a boat. A complete drilling rig may be assembled on a drilling barge, which usually is submersible; that is, it has a submersible hull or base that is flooded with water at the drilling site. Drilling equipment and crew quarters are mounted on a superstructure above the water level.

barite *n:* barium sulfate, $BaSO_4$; a mineral frequently used to increase the weight or density of drilling mud. Its specific gravity or relative density is 4.2 (i.e., it is 4.2 times heavier or denser than water). See *barium sulfate* and *mud*.

barium sulfate *n:* a chemical compound of barium, sulfur, and oxygen ($BaSO_4$). It may form a tenacious scale that is very difficult to remove. Also called barite. See *barite*.

barrel *n:* a measure of volume for petroleum products in the United States. One barrel is the equivalent of 42 U.S. gallons or 0.15899 cubic metres. One cubic metre equals 6.2897 barrels.

barrel compressor *n:* a special type of centrifugal compressor with a barrel-shaped housing.

barrel-mile *n:* a unit of measure for pipeline shipment of oil that signifies 1 barrel moved 1 mile.

barrels per day *n:* in the United States, a measure of the rate of flow of a well; the total amount of oil and other fluids produced or processed per day.

baryte *n:* variation of barite. See *barite*.

basalt *n:* an igneous rock that is dense, fine grained, and often dark gray to black in color.

base *n:* a substance capable of reacting with an acid to form a salt. A typical base is sodium hydroxide (caustic), with the chemical formula NaOH. For example, sodium hydroxide combines with hydrochloric acid to form sodium chloride (a salt) and water; this reaction is written chemically as

$$NaOH + HCl \rightarrow NaCl + H_2O.$$

basement rock *n:* either igneous or metamorphic rock, seldom containing petroleum. Ordinarily it lies below sedimentary rock. When it is encountered in drilling, the well is usually abandoned.

base metal *n:* 1. any of the reactive metals at the lower end of the electrochemical series. 2. metal to which cladding or plating is applied.

base pressure *n:* the pressure to which gas volumes are calculated, regardless of the pressure at which they are measured. The standard base pressure for gas volume calculations in the United States varies from state to state. For example, a standard cubic foot of gas in Texas is not the same as a standard cubic foot of gas in Louisiana.

basic sediment and water *n:* the water and other extraneous material present in crude oil. Usually, the BS&W content must be quite low before a pipeline will accept the oil for delivery to a refinery. The amount acceptable depends on a number of factors but usually runs from less than 5 percent to a small fraction of 1 percent.

basin *n:* a synclinal structure in the subsurface, formerly the bed of an ancient sea. Because it is composed of sedimentary rock and because its contours provide traps for petroleum, a basin is a good prospect for exploration. For example, the Permian Basin in West Texas is a major oil producer.

basket sub *n:* a fishing accessory run above a bit or a mill to recover small, nondrillable pieces of metal or junk in a well.

batch *n:* 1. a specific quantity of material that is processed, treated, or used in one operation. 2. in corrosion control, a quantity of chemical corrosion inhibitors injected into the lines of a production system. 3. in oilwell cementing, a part of the total quantity of cement to be used in a well. 4. in pipelining, a quantity of one weight of crude pumped next to one of different weight to help prevent mixing of deliveries.

batch cementing *n:* in oilwell cementing, the pumping of cement in partial amounts, or batches, as contrasted with pumping it all in one operation.

batching *n:* in pipelining, the pumping of a quantity of crude of one weight next to one of different weight to help prevent mixing of deliveries.

batching sphere *n:* a large rubber ball placed in a pipeline to separate batches. See *batch*.

batch treating *n:* the process by which a single quantity of crude oil emulsion is broken into oil and water. The emulsion is gathered and stored in a tank or container prior to treating. Compare *flow line treating*.

batch treatment *n:* in corrosion control, the injection of a quantity of chemical corrosion inhibitors into the lines of a production system, usually on a regular schedule.

bath *n:* liquid placed in a container and held at a controlled temperature to regulate the temperature of any system placed in it or passing through it.

battery *n:* 1. an installation of identical or nearly identical pieces of equipment (such as a tank battery or a battery of meters). 2. an electricity storage device.

Baumé gravity *n:* specific gravity as measured by the Baumé scale. Two arbitrary scales are employed: one for liquids lighter than water and the other for liquids heavier than water. This scale is also used to describe the density of acid solutions.

bbl *abbr:* barrel.

bbl/d *abbr:* barrels per day.

Bcf *abbr:* billion cubic feet.

Bcf/d *abbr:* billion cubic feet per day.

b/d *abbr:* barrels per day; often used in drilling reports.

B/D *abbr:* barrels per day.

BDC *abbr:* bottom dead center.

beam *n:* the extreme width (breadth) of the hull of a ship or mobile offshore drilling rig.

beam counterbalance *n:* the weights on a beam pumping unit installed on the end of the walking beam, which is opposite from the end over the well. The counterbalance offsets, or balances, the weight of sucker rods and other downhole equipment installed in the well. See *sucker rod pumping*.

beam pumping unit *n:* a machine designed specifically for sucker rod pumping, using a horizontal member (walking beam) that is worked up and down by a rotating crank to produce reciprocating motion.

bean *n:* a nipple or restriction placed in a line (a pipe) to reduce the rate of flow of fluid through the line. Beans are frequently placed in Christmas trees to regulate the flow of fluids coming out of the well. See *Christmas tree*.

bearing *n:* 1. an object, surface, or point that supports. 2. a machine part in which another part (such as a journal or pin) turns or slides.

bearing cap *n:* a device that is fitted around a bearing in order to hold or immobilize the bearing.

bearing pin *n:* a machined extension around which are placed bit bearings.

bed *n:* a specific layer of earth or rock, presenting a contrast to other layers of different material lying above, below, or adjacent to it.

bedding plane *n:* the surface that separates each successive layer of a stratified rock from its preceding layer.

bell hole *n:* a hole shaped like a bell, larger at the top than at the bottom. A bell hole may be dug beneath a pipeline to allow access for workers and tools.

bell nipple *n:* a short length of pipe (a nipple) installed on top of the blowout preventer. The top end of the nipple is expanded, or belled, to guide drill tools into the hole and usually has side connections for the fill line and mud return line.

bellows meter *n:* See *orifice meter*.

belt *n:* a flexible band or cord connecting and wrapping around each of two or more pulleys to transmit power or impart motion.

belt guard *n:* a protective grille or cover for a belt and pulleys.

bends *n:* a highly painful condition, so named because the bending joints of the body are most often affected. Also called decompression sickness. See *decompression sickness*.

bent housing *n:* a special housing for the positive-displacement downhole mud motor that is manufactured with a bend of 1-3 degrees to facilitate directional drilling.

bentonite *n:* a colloidal clay, composed primarily of montmorillonite, that swells when wet. Because of its gel-forming properties, bentonite is a major component of drilling muds. See *gel* and *mud*.

bent sub *n:* a short cylindrical device installed in the drill stem between the bottommost drill collar and a downhole mud motor. The purpose of the bent sub is to deflect the mud motor off vertical to drill a directional hole. See *drill stem*.

beta particle *n:* one of the extremely small particles, sometimes called rays, emitted from the nucleus of a radioactive substance such as radium or uranium as it disintegrates. Beta particles have a negative charge.

bevel gear *n:* one of a pair of toothed wheels whose working surfaces are inclined to nonparallel axes.

BFPH *abbr:* barrels of fluid per hour; used in drilling reports.

BHA *abbr:* bottomhole assembly.

BHP *abbr:* bottomhole pressure.

BHT *abbr:* bottomhole temperature.

billet *n:* a solid steel cylinder used to produce seamless casing. The billet is pierced lengthwise to form a hollow tube that is shaped and sized to produce the casing.

bill of lading *n:* a document by which the master of a ship acknowledges having received in good order and condition (or the reverse) certain specified goods consigned to him by some particular shipper, and binds himself to deliver them in similar condition, unless the perils of the sea, fire, or enemies prevent him, to the consignees of the shippers at the point of destination on their paying him the stipulated freight.

bioherm *n:* a reef or mound built by small organisms such as coral or oysters. Buried bioherms, formed in the geologic past, sometimes yield petroleum.

birdcage *v:* to flatten and spread the strands of a wire rope. *n:* a wire rope in such a condition.

birdcaged wire *n:* wire rope used for hoisting that has had its wires distorted into the shape of a birdcage by a sudden release of load.

bit *n:* the cutting or boring element used in drilling oil and gas wells. The bit consists of a cutting element and a circulating element. The circulating element permits the passage of drilling fluid and utilizes the hydraulic force of the fluid stream to improve drilling rates. In rotary drilling, several drill collars are joined to the bottom end of the drill pipe column, and the bit is attached to the end of the string of drill collars. Most bits used in rotary drilling are roller cone bits, but diamond bits are also used extensively. See *roller cone bit* and *diamond bit*.

bit breaker *n:* a heavy plate that fits in the rotary table and holds the drill bit while it is being made up in or broken out of the drill stem. See *bit*.

bit cone *n:* See *roller cone bit*.

bit dresser *n:* 1. a member of a cable-tool drilling crew who repairs bits. 2. a machine used to repair, sharpen, and gauge bits.

bit gauge *n:* a circular ring used to determine whether a bit is of the correct outside diameter. Bit gauges are often used to determine whether the bit has been worn down to a diameter smaller than specifications allow; such a bit is described as undergauge.

bit matrix *n:* on a diamond bit, the material (usually powdered and fused tungsten carbide) into which the diamonds are set.

bit record *n:* a report that lists each bit used during a drilling operation, giving the type of each, the footage it drilled, and the formation it penetrated.

bit shank *n:* the threaded portion of the top of the bit that is screwed into the drill collar; also called the pin.

bit sub *n:* a sub inserted between the drill collar and the bit. See *sub*.

bitumastic material *n:* a compound of asphalt and filler that is used to coat metals exposed to corrosion or weathering.

bitumen *n:* substance of dark to black color consisting almost entirely of carbon and hydrogen with very little oxygen, nitrogen, or sulfur. Bitumens occur naturally, and they can also be obtained by chemical decomposition.

bituminous shale *n:* See *oil shale*.

bl *abbr:* black; used in drilling reports.

Black Magic *n:* a proprietary name for a basic concentrate of oil-base mud. See *oil-base mud*.

blank casing *n:* casing without perforations.

blanket gas *n:* a gas phase above a liquid phase in a vessel. It is placed there for protecting the liquid from contamination, for reducing the hazard of detonation, or for pressuring the liquid. The gas has a source outside of the vessel.

blank flange *n:* a solid disk used to dead-end, or close off, a companion flange.

blank liner *n:* a liner with no perforations.

blank off *v:* to close off (as with a blank flange or bull plug).

blank pipe *n:* a pipe, usually casing, with no perforations.

blasthole drilling *n:* the drilling of holes into the earth for the purpose of placing a blasting charge (such as dynamite) in them.

bld *abbr:* bailed; used in drilling reports.

bleed *v:* to drain off liquid or gas, generally slowly, through a valve called a bleeder. To bleed down, or bleed off, means to slowly release pressure from a well or from pressurized equipment.

blind *v:* to close a line to prevent flow.

blind drilling *n:* a drilling operation in which the drilling fluid is not returned to the surface. Sometimes blind-drilling techniques are resorted to when lost circulation occurs.

blind ram *n:* an integral part of a blowout preventer that serves as the closing element on an open hole. Its ends do not fit around the drill pipe but seal against each other and shut off the space below completely. See *ram*.

blind ram preventer *n:* a blowout preventer in which blind rams are the closing elements. See *blind ram*.

block *n:* any assembly of pulleys on a common framework; in mechanics, one or more pulleys, or sheaves, mounted to rotate on a common axis. The crown block is an assembly of sheaves mounted on beams at the top of the derrick. The drilling line is reeved over the sheaves of the crown block alternately with the sheaves of the traveling block, which is raised and lowered in the derrick by the drilling line. When elevators are attached to a hook on the traveling block and drill pipe latched in the elevators, the pipe can be raised or lowered. See *crown block* and *traveling block*.

blooey line *n:* the discharge pipe from a well being drilled by air drilling. The blooey line is used to conduct the air or gas used for circulation away from the rig to reduce the fire hazard as well as to transport the cuttings a suitable distance from the well. See *air drilling*.

blowby *n:* the percentage of gases that escape past the piston rings from the combustion chamber into the crankcase of an engine.

blow case *n:* 1. a pumping device capable of transferring liquid; used to transfer crude oil and water mixtures if pump agitation would create unwanted emulsions. 2. a small tank in which liquids are accumulated and drained by applying gas or air pressure above the liquid level. Such a vessel is usually located below a pipeline or other equipment at a location where an outside power source is not convenient for removing the drained liquids. Sometimes referred to as a drip.

blowdown *n:* 1. the emptying or depressurizing of material in a vessel. 2. the material thus discarded.

blowoff cock n: a device that permits or arrests a flow of liquid from a receptacle or through a pipe, faucet, tap, or stop valve.

blowout n: an uncontrolled flow of gas, oil, or other well fluids into the atmosphere. A blowout, or gusher, can occur when formation pressure exceeds the pressure applied to it by the column of drilling fluid. A kick warns of an impending blowout. See *formation pressure, gusher,* and *kick*.

blowout preventer n: one of several valves installed at the wellhead to prevent the escape of pressure either in the annular space between the casing and drill pipe or in open hole (i.e., hole with no drill pipe) during drilling completion operations. Blowout preventers on land rigs are located beneath the rig at the land's surface; on jackup or platform rigs, at the water's surface; and on floating offshore rigs, on the seafloor. See *annular blowout preventer, inside blowout preventer,* and *ram blowout preventer*.

blowout preventer control panel n: a set of controls, usually located near the driller's position on the rig floor, that is manipulated to open and close the blowout preventers. See *blowout preventers*.

blowout preventer control unit n: a service that stores hydraulic fluid under pressure in special containers and provides a method to open and close the blowout preventers quickly and reliably. Usually, compressed air and hydraulic pressure provide the opening and closing force in the unit. See *blowout preventer*.

BLPD abbr: barrels of liquid per day, usually used in reference to total production of oil and water from a well.

BO abbr: barrels of oil; used in drilling reports.

boilaway test n: also called weathering test. See *weathering test*.

boiler n: a closed pressure vessel that has a furnace equipped to burn coal, oil, or gas and is used to generate steam from water.

boiler house v: (slang) to make up or fake a report.

boiling point n: the temperature at which the vapor pressure of a liquid becomes equal to the pressure exerted on the liquid by the surrounding atmosphere. The boiling point of water is 212°F or 100°C at atmospheric pressure (14.7 psig or 101.325 kPa).

boll weevil n: (slang) an inexperienced rig or oil field worker; sometimes shortened to *weevil*.

boll weevil corner n: (slang, obsolete) the work station of an inexperienced rotary helper, on the opposite side of the rotary from the pipe racker.

boll weevil hanger n: a tubing hanger.

bomb n: a thick-walled container, usually steel, used to hold samples of oil or gas under pressure. See *bottomhole pressure*.

bond n: the adhering or joining together of two materials (as cement to formation). v: to adhere or to join to another material.

boom n: a movable arm of tubular or bar steel, used on some types of cranes or derricks to support the hoisting lines that carry the load.

boom dog n: a ratchet device on a crane that prevents the boom of the crane from being lowered but still permits it to be raised. It is also called a boom ratchet.

boomer n: 1. (slang) an oil field worker who moves from one center of activity to another; a floater or transient. 2. a device used to tighten chains on a load of pipe or other equipment on a truck to make it secure.

boom ratchet n: See *boom dog*.

boom stop n: the steel projections on a crane struck by the boom if it is raised too high or lowered too far.

booster station n: an installation on a pipeline that maintains or increases pressure of the fluid coming through the pipeline and being sent on to the next station or terminal.

boot n: a tubular device placed in a vertical position, either inside or outside a larger vessel, through which well fluids are conducted before they enter the larger vessel. A boot aids in the separation of gas from wet oil. Also called a flume or conductor pipe.

boot basket n: See *junk sub*.

boot sub n: See *junk sub*.

BOP abbr: blowout preventer.

BOPD abbr: barrels of oil per day.

BOP stack n: the assembly of blowout preventers installed on a well.

bore n: 1. the inside diameter of a pipe or a drilled hole. 2. the diameter of the cylinder of an engine.

borehole n: the wellbore; the hole made by drilling or boring. See *wellbore*.

bottled gas n: liquefied petroleum gas placed in small containers for sale to domestic customers.

bottleneck n: an area of reduced diameter in pipe, brought about by excessive longitudinal strain or by a combination of longitudinal strain and the swagging action of a body. A bottleneck may result if the downward motion of the drill pipe is stopped with the slips instead of the brake.

bottle test n: a test in which varying amounts of a chemical are added to bottled samples of an emulsion to determine how much of the chemical is needed to break the emulsion into oil and water.

bottom dead center n: the position of the piston at the lowest point possible in the cylinder of an engine. It is often marked on the engine flywheel.

bottom hold-down n: a mechanism for anchoring a bottomhole pump in a well, located on the lower end of the pump. Compare *top hold-down*.

bottomhole n: the lowest or deepest part of a well. adj: pertaining to the bottom of the wellbore.

bottomhole assembly n: the portion of the drilling assembly below the drill pipe. It can be very simple – composed of only the bit and drill collars – or it can be very complex and made up of several drilling tools.

bottomhole choke n: a device with a restricted opening placed in the lower end of the tubing to control the rate of flow. See *choke*.

bottomhole contract n: a contract providing for the payment of money or other considerations upon the completion of a well to a specified depth.

bottomhole packer n: a device that blocks passage through the annular space between two strings of pipe and is installed near the bottom of the hole. See *packer*.

bottomhole plug n: a bridge plug or cement plug placed near the bottom of the hole to shut off a depleted, water-producing, or unproductive zone.

bottomhole pressure n: 1. the pressure at the bottom of a borehole. It is caused by the hydrostatic pressure of the drilling fluid in the hole and, sometimes, any back-pressure held at the surface, as when the well is shut in with blowout preventers. When mud is being circulated, bottomhole pressure is the hydrostatic pressure plus the remaining circulating pressure required to move the mud up the annulus. 2. the pressure in a well at a point opposite the producing formation, as recorded by a bottomhole pressure bomb. See *bottomhole pressure bomb*.

bottomhole pressure bomb n: a bomb used to record the pressure in a well at a point opposite the producing formation. See *bomb*.

bottomhole pressure gauge n: a gauge to measure bottomhole pressure. See *bottomhole pressure*.

bottomhole pump n: any of the rod pumps, high-pressure liquid pumps, or centrifugal pumps located at or near the bottom of the well and used to lift the well fluids. See *sucker rod pumping*, *hydraulic pumping*, and *submersible pump*.

bottomhole separator n: a device used to separate oil and gas at the bottom of wells to increase the volumetric efficiency of the pumping equipment.

bottomhole temperature n: temperature measured in a well at a depth at the midpoint of the thickness of the producing zone.

bottom loading pressure n: the pressure exerted on the bottom hull of a column-stabilized, semisubmersible drilling rig when the rig is submerged.

bottoms n: 1. the liquids and the residue that collect in the bottom of a vessel (such as tank bottoms) or that remain in the bottom of a storage tank after a period of service. 2. the residual fractions remaining at the bottom of a fractionating tower after lighter components have been distilled off as vapors.

bottoms up n: a complete trip from the bottom of the wellbore to the top.

bottom time n: the total amount of time, measured in minutes, from the time a diver leaves the surface until he begins his ascent.

bottom water n: water found below oil and gas in a producing formation.

bottom wiper plug n: a device placed in the cementing head and run down the casing in front of cement to clean the mud off the walls of the casing and to prevent contamination between the mud and the cement.

bounce dive n: a rapid dive with a very short bottom time to minimize decompression time.

Bourdon tube n: a flattened metal tube bent in a curve, which tends to straighten when pressure is applied internally. By the movements of an indicator over a circular scale, a Bourdon tube indicates the pressure applied.

bowl n: an insert that fits into the opening of a master bushing and accommodates the slips. See *insert*.

bowline knot n: a knot used primarily in lifting heavy equipment with the catline, since it can be readily tied and untied regardless of the weight of the load on it.

bow lines *n pl:* the lines running from the bow of a mobile offshore drilling rig, especially the forward mooring lines.

box *n:* the female section of a connection. See *tool joint.*

box and pin *n:* See *tool joint.*

box tap *n:* old-style tap with longitudinal grooves across the threads. See *tap* and *taper tap.*

box threads *n:* threads on the female section, or box, of a tool joint. See *tool joint.*

Boyle's law *n:* a gas law that concerns pressure. It states that for any ideal gas or mixture of ideal gases at any definite temperature, the product of the absolute pressure times the volume is a constant ($PV = K$).

bpd or BPD *abbr:* barrels per day.

BPH *abbr:* barrels per hour; used in drilling reports.

B-P mix *n:* a liquefied hydrocarbon product composed chiefly of butanes and propane. If it originates from a refinery, it may also contain butylenes and propylene.

brackish water *n:* water that contains relatively low concentrations of soluble salts. Brackish water is saltier than fresh water but not as salty as salt water.

bradding *n:* a condition in which the weight on a bit tooth has been so great that the tooth has dulled until the softer inner portion of the tooth caves over the harder case area.

bradenhead *n:* (obsolete) casinghead.

bradenhead flange *n:* a flanged connection at the top of the oilwell casing.

bradenhead squeezing *n:* the process by which hydraulic pressure is applied to a well to force fluid or cement outside the wellbore without the use of a packer. The bradenhead, or casinghead, is closed to shut off the annulus when making a bradenhead squeeze. Although this term is still used, the term *bradenhead* is obsolete. See *annular space, casinghead,* and *squeeze.*

brake *n:* a device for arresting the motion of a mechanism, usually by means of friction, as in the drawworks brake. Compare *electrodynamic brake* and *hydrodynamic brake.*

brake band *n:* a part of the brake mechanism, consisting of a flexible steel band lined with asbestos or a similar material, that grips a drum when tightened. On a drilling rig, the brake band acts on the flanges of the drawworks drum to control the lowering of the traveling block and its load of drill pipe, casing, or tubing.

brake block *n:* a section of the lining of a band brake; it is shaped to conform to the curvature of the band and is attached to it with countersunk screws. See *brake band.*

brake flange *n:* the surface on a winch, drum, or reel where the brake is applied to control the movement of the unit through friction.

brake lining *n:* the part of the brake that presses on the brake drum. On the drawworks, the circular series of brake blocks are bolted to the brake bands with countersunk brass bolts; the lining is the frictional, or gripping, element of a mechanical brake.

brake rider *n:* (slang) a driller who is said to rely too heavily on the drawworks brake.

brake rim: *n:* also called brake flange. See *brake flange.*

branch line *n:* a line, usually a pipe, joined to and diverging from another line.

brass running nipple *n:* a device used in the flow cross of the Christmas tree as a thread protector while the rods are being run. Because it is brass, it prevents friction sparks.

breadth *n:* the greatest overall dimension measured perpendicular to the longitudinal centerline of the hull of a mobile offshore drilling rig. Also called the beam. See *beam.*

break *v:* to begin or start (as, to break circulation or break tour.)

break circulation *v:* to start the mud pump for restoring circulation of the mud column. Because the stagnant drilling fluid has thickened or gelled during the period of no circulation, a high pump pressure is usually required to break circulation.

breakdown *n:* a failure of equipment. *adj:* pertaining to the amount of pressure needed at the wellhead to rupture the formation in a fracture treatment or squeeze job (as formation breakdown pressure).

breaking down *v:* unscrewing the drill stem into single joints and placing them on the pipe rack. The operation takes place upon completion of the well, or in changing from one size of pipe to another. See *lay down pipe.*

breaking strength *n:* the load under which a chain or a rope breaks.

break out *v:* 1. to unscrew one section of pipe from another section, especially drill pipe while it is being withdrawn from the wellbore. During this operation, the tongs are used to start the unscrewing operation. 2. to separate, as gas from a liquid or water from an emulsion.

breakout block *n:* a heavy plate that fits in the rotary table and holds the drill bit while it is being unscrewed from the drill collar. See *bit breaker*.

breakout cathead *n:* a device, attached to the catshaft of the drawworks, that is used as a power source for unscrewing drill pipe; usually located opposite the driller's side of the drawworks. See *cathead*.

breakout tongs *n:* tongs that are used to start unscrewing one section of pipe from another section, especially drill pipe coming out of the hole. See *tongs* and *lead tongs*.

breakover *n:* the change in the chemistry of a mud from one type to another; also called a conversion.

break tour *v:* to begin operating 24 hours per day. Moving the rig and rigging up are usually carried on during daylight hours only. When the rig is ready for operation at a new location, crews break tour and start operating 24 hours per day. See *tour*.

breathe *v:* to move with a slight, regular rhythm. Breathing occurs in tanks of vessels when vapors are expelled and air is taken in. For example, a tank of crude oil expands because of the rise in temperature during the day and contracts as it cools at night, expelling vapors as it expands and taking in air as it contracts. Tubing breathes when it moves up and down in sequence with a sucker rod pump.

breather *n:* a small vent in an otherwise airtight enclosure for maintaining equality of pressure within and without.

breathing bag *n:* part of the semiclosed circuit breathing apparatus, used to mix gas and ensure low breathing resistance.

bridge *n:* 1. an obstruction in the borehole, usually caused by the caving in of the wall of the borehole or by the intrusion of a large boulder. 2. a tool placed in the hole to retain cement or other material that may later be removed, drilled out, or left permanently.

bridge plug *n:* a downhole tool, composed primarily of slips, a plug mandrel, and a rubber sealing element, that is run and set in casing to isolate a lower zone while an upper section is being tested or cemented.

bridging material *n:* the fibrous, flaky, or granular material added to cement slurry or drilling fluid to aid in sealing formations in which lost circulation has occurred. See *lost circulation* and *lost circulation material*.

bridle *n:* a cable on a pumping unit, looped over the horse head and connected to the carrier bar to support the polished-rod clamp. See *sucker rod pumping*.

brine *n:* water that has a large quantity of salt, especially sodium chloride, dissolved in it; salt water.

bring in a well *v:* to complete a well and put it on producing status.

British thermal unit *n:* a measure of heat energy equivalent to the amount of head needed to raise 1 pound of water 1 degree Fahrenheit.

brkn *abbr:* broken; used in drilling reports.

Brownian movement *n:* the random movement exhibited by microscopic particles when suspended in liquids or gases, caused by the impact of molecules of fluid surrounding the particle.

BS&W *abbr:* basic sediment and water.

BS&W monitor *n:* a device used in LACT systems to measure and record automatically the amount of water and other contaminants in oil being transferred to a pipeline and to divert contaminated oil to treatment facilities.

Bscf/d *abbr:* billion standard cubic feet per day.

Btu *abbr:* British thermal unit.

bubble cap *n:* a metal cap, mounted on a tray, that has openings allowing vapor bubbles in a gas-processing tower to contact cool liquids, causing some of the vapor to condense to liquid.

bubble-cap tray *n:* a perforated steel tray on which bubble caps are mounted. Bubble caps and trays are arranged in bubble towers, cylindrical vessels set vertically. See *sieve tray* and *valve tray*.

bubble point *n:* 1. the temperature and pressure at which part of a liquid begins to convert to gas. For example, if a certain volume of liquid is held at constant pressure, but its temperature is increased, a point is reached when bubbles of gas begin to form in the liquid. That is the bubble point. Similarly, if a certain volume of liquid is held at a constant temperature but the pressure is reduced, the point at which gas begins to form is the bubble point. Compare *dew point*. 2. the temperature and pressure at which gas, held in solution in crude oil, breaks out of solution as free gas.

bubble tower *n*: a vertical cylindrical vessel in which bubble caps and bubble-cap trays are arranged.

bubble tray *n*: See *bubble-cap tray*.

buckling stress *n*: bending of the pipe that may occur due to deviation of the hole. The pipe may bend because of the angle of the hole or because of an abrupt deviation such as a dog leg.

buck up *v*: to tighten up a threaded connection (such as two joints of drill pipe).

buddy system *n*: a method of pairing two individuals for their mutual aid or protection. The buddy system is used among crew members to ensure that each man is accounted for, particularly when running a test where hydrogen sulfide may be encountered.

bug blower *n*: (slang) a large fan installed on a drilling rig to blow insects away from the work area; any type of fan.

buildup test *n*: a test in which a well is shut in for a prescribed period of time and a bottomhole pressure bomb run in the well to record the pressure. From this data and from knowledge of pressures in a nearby well, the effective drainage radius or the presence of permeability barriers or other production deterrents surrounding the wellbore can be estimated.

bulkhead *n*: an interior wall that subdivides a ship or a mobile offshore drilling rig into compartments.

bulkhead deck *n*: the highest deck to which water bulkheads extend on a ship or a mobile offshore drilling rig.

bulk plant *n*: a wholesale distributing point for products made from natural gas and petroleum.

bulldozer *n*: a powerful tractor having a vertical blade at the front end for moving rocks, dirt, and like materials.

bullet perforator *n*: a tubular device that, when lowered to a selected depth within a well, fires bullets through the casing to provide holes through which the formation fluids may enter the wellbore.

bullets *n pl*: the devices loaded into perforating guns to penetrate into casing and cement and for some distance into the formation when the guns are fired. See *gun-perforate*.

bull gear *n*: the large circular gear in a mud pump that is driven by the prime mover and in turn drives the connecting rods.

bull plug *n*: a threaded nipple with a rounded closed end, used to stop up a hole or close off the end of a line.

bump a well *v*: to cause the pump on a pumping unit to hit the bottom of the well by having too long a sucker rod string in the unit.

bumped *adj*: in cementing operations, pertaining to a cement plug which comes to rest on the float collar. A cementing operator may say, "I have a bumped plug," when the plug strikes the float collar.

bumper jar *n*: an expansion joint, permitting vertical movement of the upper section without movement of the lower section of the tool, used to deliver a heavy blow to objects in the borehole. If a fish can be freed by a downward blow, a bumper jar can be very effective. See *fish*.

bumper sub *n*: a device similar to a jar but used in the drill stem to compensate for vertical movement of the stem, especially in offshore drilling. It also provides jarring action, but to a lesser extent than a jar. See *jar*.

bump off a well *v*: to disconnect a pull-rod line from a central power unit.

bunker C oil *n*: See *residuals*.

bunkers *n*: heavy fuel oil (#6 oil) used by ships as fuel to power the vessel.

bunkhouse *n*: a building providing sleeping quarters for workers.

buoyancy *n*: the apparent loss of weight of an object immersed in a fluid. If the object is floating, the immersed portion displaces a volume of fluid the weight of which is equal to the weight of the object.

burning point *n*: the lowest temperature at which an oil or fuel will burn when an open flame is held near its surface.

burn pit *n*: an earthen pit in which waste oil and other materials are burned.

burn shoe *n*: a type of rotary shoe designed to mill away tubular goods (such as casing, drill pipe, and so on) causing stuck pipe; used in fishing operations.

burst pressure *n*: the internal pressure stress on casing or other pipe. Burst pressure occurs when the pipe's internal pressure is greater than its external pressure, causing the pipe to burst.

burst strength *n*: the internal pressure that can cause pipe to rupture.

bus *n*: an assembly of electrical conductors for collecting current from several sources and distributing it to feeder lines so that it will be available where needed. Also called bus bar.

bus bar *n:* also called bus. See *bus.*

bushing *n:* 1. a pipe fitting on which the external thread is larger than the internal thread to allow two pipes of different sizes to be connected together. 2. a removable lining or sleeve inserted or screwed into an opening to limit its size, resist wear or corrosion, or serve as a guide.

butane *n:* a paraffin hydrocarbon, C_4H_{10}, that is a gas in atmospheric conditions but is easily liquefied under pressure; a constituent of liquefied petroleum gas. See *commercial butane, field grade butane,* and *normal butane.*

butane, commercial *n:* See *commercial butane.*

butene *n:* See *butylene.*

Butterworth tank cleaning system *n:* trade name for apparatus for cleaning and freeing oil tanks of gas by means of high-pressure jets of hot water. It consists essentially of opposed double nozzles, which rotate slowly about their horizontal and vertical axes, projecting two streams of hot water at a pressure of 175 psi against all inside surfaces of the deck, bulkheads, and shell plating.

button bit *n:* a drilling bit with tungsten carbide inserts that resemble buttons. See *roller cone bit.*

butylene *n:* hydrocarbon member of the olefin series having the chemical formula C_4H_8. Official name is butene.

BW *abbr:* barrels of water; used in drilling reports.

BWPD *abbr:* barrels of water per day.

BWPH *abbr:* barrels of water per hour; used in drilling reports.

bypass *n:* a pipe connection around a valve or other control mechanism, installed to permit passage of fluid through the line while adjustments or repairs are being made on the control.

bypass valve *n:* a valve that permits flow around a control valve, a piece of equipment, or a system.

C *abbr:* Celsius (formerly centigrade). See *Celsius scale*.

C *sym:* coulomb.

cable *n:* a rope of wire, hemp, or other strong fibers. See *wire rope*.

cable-tool drilling *n:* a drilling method in which the hole is drilled by dropping a sharply pointed bit on the bottom of the hole. The bit is attached to a cable, and the cable is picked up and dropped, picked up and dropped, over and over, as the hole is drilled.

cage *n:* in a sucker rod pump, the device that contains and confines the valve ball and keeps it within the proper operating distance from the valve seats.

cage wrench *n:* a special wrench designed for use in connecting the cage of a sucker rod pump to the sucker rod string.

cake thickness *n:* the thickness of drilling mud filter cake. See *filter cake*.

calcium carbonate *n:* a chemical combination of calcium, carbon, and oxygen; the main constituent of limestone. It forms a tenacious scale in water-handling facilities and is a cause of water hardness. Chemical formula is $CaCO_3$.

calcium chloride *n:* a moisture-absorbing chemical compound, or desiccant, used as an accelerator in cements and as a drying agent. Its formula is $CaCl_2$.

calcium sulfate *n:* a chemical compound of calcium, sulfur, and oxygen. Its formula is $CaSO_4$.

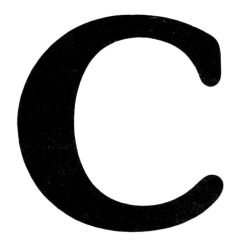

Although sometimes considered a contaminant of drilling fluids, it may at times be added to them to produce certain properties. Like calcium carbonate, it forms scales on water-handling facilities, which may be hard to remove. See *anhydrite* and *gypsum*.

calcium-treated mud *n:* a freshwater drilling mud using calcium oxide (lime) or calcium sulfate (gyp) to retard the hydrating qualities of shale and clay formations, thus facilitating drilling. Calcium-treated muds resist salt and anhydrite contamination but may require further treatment to prevent gelation (solidification) under the high temperatures of deep wells.

calibration *n:* the adjustment or standardizing of a measuring instrument. Log calibration is based on

Crown block

the use of a permanent calibration facility of the American Petroleum Institute at the University of Houston to establish standard units for nuclear logs.

calibration tank n: See prover tank.

caliper log n: a record whereby the diameter of the wellbore is ascertained, indicating undue enlargement due to caving in, washout, or other causes. The caliper log also reveals corrosion, scaling, or pitting inside tubular goods.

calorie n: the amount of heat energy necessary to raise the temperature of 1 gram of water 1 degree Celsius. It is the metric equivalent of the British thermal unit.

calorimeter n: an apparatus used to determine the heating value of a combustible material.

cam n: an eccentrically shaped disk, mounted on a camshaft, that varies in distance from its center to various points on its circumference. As the camshaft is rotated, a set amount of motion is imparted to a follower riding on the surface of the cam. In the internal-combustion engine, cams are used to operate the intake and the exhaust valves.

cam follower n: output link of a cam mechanism. See cam.

camshaft n: the cylindrical bar used to support a rotating device called a cam. See cam.

Canadian Association of Oilwell Drilling Contractors n: a trade association whose head office is located at 614 Manulife House, 603 - 7th Avenue S.W., Calgary, Alberta T2P 2T5. The association represents virtually 100 percent of the rotary drilling contractors and the majority of well service rig operators in western Canada. The organization concerns itself with research, education, accident prevention, government relations, and other matters of interest to members.

candela n: the fundamental unit of luminous intensity in the metric system. The symbol for candela is cd.

canted leg adj: pertaining to an independent-leg jackup rig designed so that the legs may be slanted outward to increase support against lateral stresses when the unit is on the seafloor.

canvas packer n: (obsolete) a device for sealing the annular space between the top of a liner and the existing casing string.

CAODC abbr: Canadian Association of Oilwell Drilling Contractors.

capacitance probe n: a device used in most net-oil computers that senses the different dielectric constants of oil and water in a water-oil emulsion. See dielectric constant.

capacitor n: an electrical device that, when wired in the line of an electrical circuit, stores a charge of electricity and returns the charge to the line when certain electrical conditions occur. Also called a condenser. See condenser.

cap a well v: to control a blowout by placing a very strong valve on the wellhead. See blowout.

capillaries n pl: fissures or cracks in a formation through which water or hydrocarbons flow.

capillary pressure n: a pressure or adhesive force caused by the surface tension of water. This pressure causes the water to rise higher in small capillaries in the formation than it does in large capillaries. Capillary pressure in a rock formation is comparable to the pressure of water that rises higher in a small glass capillary tube than it does in a larger tube.

cap rock n: 1. impermeable rock overlying an oil or gas reservoir that tends to prevent migration of oil or gas out of the reservoir. 2. the porous and permeable strata overlying salt domes that may serve as the reservoir rock.

capture cross section n: the tendency of elements in their compounds to reduce energy or number of particles by absorbing them. The more densely populated an area may be, the more certain it is that energy will be absorbed or that the particles will be retained within the atomic structure.

capture gamma ray n: a high-energy gamma ray emitted when the nucleus of an atom captures a neutron and becomes intensely excited. Capture gamma rays are counted by the neutron logging detector.

carbonate n: 1. a salt of carbonic acid. 2. a compound containing the carbonate (CO_3^{--}) radical.

carbonate reef n: See reef.

carbonate rock n: a sedimentary rock composed primarily of calcium carbonate (limestone) or calcium magnesium carbonate (dolomite); sometimes makes up petroleum reservoirs.

carbon black n: very fine particles of almost pure amorphous carbon, usually produced from gaseous or liquid hydrocarbons by thermal decomposition or by controlled combustion with a restricted air supply.

carbon dioxide n: a colorless, odorless gaseous compound of carbon and oxygen (CO_2). A product of combustion and a filler for fire extinguishers, this heavier-than-air gas can collect in low-lying areas, where it may displace oxygen and present the hazard of anoxia.

carbon dioxide excess *n:* also called hypercapnia. See *hypercapnia*.

carbon log *n:* a record that indicates the pressure of hydrocarbons by measuring carbon atoms and reveals the presence of water by measuring oxygen atoms. Oil and water saturations can be closely approximated without the requirement of adequate salinity of known concentration in order to calculate saturations (as in resistivity and pulsed neutron logs).

carbon monoxide *n:* a colorless, odorless gaseous compound of carbon and oxygen (CO). A product of incomplete combustion, it is extremely poisonous to breathe.

carbonyl sulfide *n:* a chemical compound of the aldehyde groups containing a carbonyl group and sulfur (COS). It is a contaminant in gas liquids, usually removed to meet sulfur specifications.

carboxymethyl cellulose *n:* a nonfermenting cellulose product used in drilling fluids to combat contamination from anhydrite (gyp) and to lower the water loss of the mud.

carrier rig *n:* a self-propelled, wheeled unit used to service oil and gas wells. Modern production rigs are usually carrier units, having the masts, hoists, engines, and other auxiliaries needed to service or work over a well mounted on a chassis powered by the engines used for hoisting. See *back-in unit* and *drive-in unit*.

cascade system *n:* in respiratory systems, a series connection of air cylinders in which the output of air from one adds to that of the next.

cased *adj:* pertaining to a wellbore in which casing has been run and cemented. See *casing*.

cased hole *n:* a wellbore in which casing has been run. See *casing*.

case-hardened *adj:* hardened (as for a ferrous alloy) so that the surface layer is harder than the interior.

casing *n:* steel pipe placed in an oil or gas well as drilling progresses to prevent the wall of the hole from caving in during drilling, to prevent seepage of fluids, and to provide a means of extracting petroleum if the well is productive.

casing adapter *n:* a swage nipple, usually beveled, installed on the top of a string of pipe that does not extend to the surface. It prevents a smaller string of pipe or tools from hanging up on the top of the column when it is run into the well.

casing burst pressure *n:* the amount of pressure that, when applied to a string of casing, causes the wall of the casing to fail. This pressure is critically important when a gas kick is being circulated out because gas on the way to the surface expands and exerts more pressure than it exerted at the bottom of the well.

casing centralizer *n:* a device secured around the casing at regular intervals to center it in the hole. Casing that is centralized allows a more uniform cement sheath to form around the pipe.

casing coupling *n:* a tubular section of pipe that is threaded inside and used to connect two joints of casing.

casing cutter *n:* a heavy cylindrical body, fitted with a set of knives, used to free a section of casing in a well. The cutter is run downhole on a string of tubing or drill pipe, and the knives are rotated against the inner walls of the casing to free the section that is stuck.

casing design chart *n:* a list of the various grades and weights of pipe available for the casing string, which includes pertinent safety factors such as stress effects to permit selection of the most economical and safest casing for a specific job.

casing elevator *n:* See *elevator*.

casing float collar *n:* See *float collar*.

casing float shoe *n:* See *float shoe*.

casing hanger *n:* a circular device with a frictional gripping arrangement, used to suspend casing in a well.

casinghead *n:* a heavy, flanged steel fitting connected to the first string of casing; it provides a housing for slips and packing assemblies, allows suspension of intermediate and production strings of casing, and supplies the means for the annulus to be sealed off. Also called a spool.

casinghead gas *n:* gas produced with oil.

casinghead gas contract *n:* a form of contract used by industry for the purchase and sale of casinghead gas.

casinghead gasoline *n:* obsolete term for natural gasoline.

casing overshot *n:* also called a casing-patch tool. See *casing-patch tool*.

casing pack *n:* a means of cementing casing in a well so that the casing may, if necessary, be retrieved with minimum difficulty. A special mud, usually an oil mud, is placed in the well ahead of the cement after the casing has been set. The mud used is nonsolidifying so that it does not bind or stick to the casing in the hole in the area above the cement. Since the mud does not gel for a long time, the casing can be cut above the cemented section and retrieved. Casing packs are used in wells of doubtful or limited production to permit reuse of valuable lengths of casing.

casing-pack v: to cement casing in a well in a way that allows for easy retrieval of sections of casing. See *casing pack*.

casing-patch tool n: a special tool with a rubber packer or lead seal that is used to repair casing. When casing is damaged downhole, a cut is made below the damaged casing, the damaged casing and the casing above it are pulled from the well, and the damaged casing is removed from the casing string. The tool is made up and lowered into the well on the casing until it engages the top of the casing that remains in the well, and a rubber packer or lead seal in the tool forms a seal with the casing that is in the well. The casing-patch tool is an overshotlike device and is sometimes called a casing overshot.

casing point n: the depth in a well at which casing is set, generally the depth at which the casing shoe rests.

casing pressure n: the pressure built up in a well between the casing and tubing or the casing and drill pipe.

casing protector n: a short threaded nipple screwed into the open end of the coupling and over the threaded end of casing to protect the threads from dirt accumulation and damage. Also called a thread protector, it is made of steel or plastic. See *thread protector*.

casing rack n: also called pipe rack. See *pipe rack*.

casing roller n: a rugged tool composed of a mandrel with a series of eccentric roll surfaces, each of which is assembled with a series of heavy-duty rollers. It is used to restore buckled, collapsed, or dented casing in a well to normal diameter and roundness. Made up on tubing or drill pipe and run into the well to the depth where the casing is deformed, the tool is rotated slowly, allowing the rollers to contact all the sides of the casing and restore it to some semblance of its original condition.

casing seat n: the location of the bottom of a string of casing that is cemented in a well; typically, a casing shoe is made up on the end of the casing at this point.

casing shoe n: also called a guide shoe. See *guide shoe*.

casing slips n: See *spider*.

casing spear n: a fishing tool designed to grab casing from the inside so that when the spear is retrieved, the attached casing comes with it.

casing spider n: See *spider*.

casing string n: the entire length of all the joints of casing run in a well. Casing is manufactured in lengths of about 30 feet (9 metres), each length or joint being joined to another as casing is run in a well. See *combination string*.

casing swage n: a solid cylindrical body, pointed at the bottom and equipped with a tool joint at the top for connection with a jar, used to make an opening in a collapsed casing and drive it back to its original shape.

casing tongs n: the large wrenches used for turning when making up or breaking out casing. See *tongs*.

cat n: a tractor designed to move easily over rough terrain, used often to clear areas in earth-moving operations and in skidding rigs. It is a shortened form of Caterpillar, a trade name.

catalyst n: a substance that alters, accelerates, or instigates chemical reactions without itself being affected.

catalytic cracking n: a type of cracking similar to thermal cracking but involving the use of catalysts and lower temperatures and pressures. It produces a gasoline that has a higher octane rating and a lower sulfur content than that produced by thermal cracking.

catch samples v: to obtain cuttings for geological information as formations are penetrated by the bit. The samples are obtained from drilling fluid as it emerges from the wellbore or, in cable-tool drilling, from the bailer. Cuttings are carefully washed until they are free of foreign matter, dried, and labeled to indicate the depth from which they were obtained.

catenary n: the curve assumed by a perfectly flexible line hanging under its own weight between two fixed points.

cathead n: a spool-shaped attachment on the end of the catshaft, around which rope for hoisting and pulling is wound. See *breakout cathead* and *makeup cathead*.

cathead spool n: also called the cathead. See *cathead*.

cathode *n:* 1. one of two electrodes in an electrolytic cell, represented as the positive terminal of a cell. 2. in cathodic protection systems, the protected structure that is representative of the cathode and is protected by having a conventional current flow from an anode to the structure through the electrolyte.

cathodic protection *n:* a method of protecting a metal structure from corrosion by making its surfaces cathodic and controlling the location of anodic areas so that corrosion damage can be reduced to tolerable levels.

cation *n:* a positively charged ion; the ion in an electrolyzed solution that migrates to the cathode. See *ion*. Compare *anion*.

catline *n:* a hoisting or pulling line powered by the *cathead* and used to lift heavy equipment on the rig. See *cathead*.

catshaft *n:* an axle that crosses through the drawworks and contains a revolving spool called a cathead at either end. See *cathead*.

catwalk *n:* 1. the ramp at the side of the drilling rig where pipe is laid out to be lifted to the derrick floor by the catline. See *catline*. 2. any elevated walkway.

caustic soda *n:* sodium hydroxide, used to maintain an alkaline pH in drilling mud and in petroleum fractions. Its formula is NaOH.

caustic treater *n:* a vessel holding sodium hydroxide or other alkalis through which a solution flows for removal of sulfides, mercaptans, or acids.

cavernous formation *n:* a rock formation that contains large open spaces, usually resulting from the dissolving of soluble substances by formation waters that may still be present. See *vug*.

caving *n:* collapsing of the walls of the wellbore; also called sloughing.

cavings *n pl:* particles that fall off (are sloughed from) the wall of the wellbore. Not the same as cuttings.

cc *abbr:* cubic centimetre.

cd *sym:* candela.

cellar *n:* a pit in the ground providing additional height between the rig floor and the wellhead to accommodate the installation of blowout preventers, rathole, mousehole, and so forth. It also collects drainage water and other fluids for subsequent disposal.

cellar deck *n:* the lower deck of a double-decked semisubmersible drilling rig. See *main deck* or *Texas deck*.

cellophane *n:* a thin transparent material made from cellulose and used as a lost circulation material. See *cementing material*.

Celsius scale *n.* the metric scale of temperature measurement used universally by scientists. On this scale, 0 degrees represents the freezing point of water and 100 degrees its boiling point at a barometric pressure of 760 mm. Degrees Celsius are converted to degrees Fahrenheit by using the following equation:

$$°F = 9/5 \, (°C) + 32.$$

The Celsius scale was formerly called the centrigrade scale; now, however, the term *Celsius* is preferred in the International System of Units (SI).

cement *n:* a powder, consisting of alumina, silica, lime, and other substances that hardens when mixed with water. Extensively used in the oil industry to bond casing to the walls of the wellbore.

cement additive *n:* a material added to cement to change its properties. Chemical accelerators, chemical retarders, and weight-reduction materials are common additives. See *cementing materials*.

cement bond *n:* the adherence of casing to cement and cement to formation. When casing is run in a well, it is set, or bonded, to the formation by means of cement.

cement bond survey *n:* an acoustic survey or sonic-logging method that records the quality or hardness of the cement used in the annulus to bond the casing and the formation. Casing that is well bonded to the formation transmits an acoustic signal quickly; poorly bonded casing transmits a signal slowly. See *acoustic survey* and *sonic logging*.

cement-casing *v:* filling the annulus beween the casing and wall of the hole with cement to support the casing and prevent fluid migration between permeable zones.

cement channeling *n:* an undesirable phenomenon that can occur when casing is being cemented in a borehole. The cement slurry fails to rise uniformly between the casing and the borehole wall, leaving spaces devoid of cement. Ideally, the cement should completely and uniformly surround the casing and form a strong bond to the borehole wall.

cement clinker *n:* a substance formed by melting ground limestone, clay or shale, and iron ore in a kiln. Cement clinker is ground into a powdery mixture and combined with small amounts of gypsum or other materials to form cement.

cement dump bailer n: a cylindrical container with a valve, used to release small batches of cement in a remedial cementing operation.

cementing n: the application of a liquid slurry of cement and water to various points inside or outside the casing. See *primary cementing, secondary cementing,* and *squeeze cementing.*

cementing barge n: a barge containing the cementing pumps and other equipment needed for oilwell cementing in water operations.

cementing basket n: a collapsible or folding metal cone that fits against the walls of the wellbore to prevent the passage of cement; sometimes called a metal-petal basket.

cementing head n: an accessory attached to the top of the casing to facilitate cementing of the casing. It has passages for cement slurry and retainer chambers for cementing wiper plugs.

cementing materials n pl: a slurry of portland cement and water and sometimes one or more additives that affect either the density of the mixture or its setting time. The portland cement used may be high early strength, common (standard), or slow setting. Additives include accelerators (such as calcium choloride), retarders (such as gypsum), weighting materials (such as barium sulfate), lightweight additives (such as bentonite), and a variety of lost circulation materials (such as mica flakes).

cementing pump n: a high-pressure pump used to force cement down the casing and into the annular space between the casing and the wall of the borehole.

cementing time n: the total elapsed time needed to complete a cementing operation.

cement plug n: a portion of cement placed at some point in the wellbore to seal it. See *cementing.*

cement retainer n: a tool set temporarily in the casing or well to prevent the passage of cement, thereby forcing it to follow another designated path. It is used in squeeze cementing and other remedial cementing jobs.

cement system n: a particular slurry containing cement and water, with or without additives.

centerline n: the middle line of the hull of a mobile offshore drilling rig from stem to stern, as shown in a waterline view.

center of buoyancy n: the center of gravity of the fluid displaced by a floating body (such as a ship or mobile offshore drilling rig).

center of flotation n: the geometric center of the water plane at which a mobile offshore drilling rig floats and about which a rig rotates when acted upon by an external force without a change in displacement.

center of gravity n: the point at which an object can be supported so that it balances, and at which all gravitational forces on the body and the weight of the body are concentrated; the center of mass.

center of pressure n: the point at which all wind pressure forces are concentrated.

centigrade scale n: See *Celsius scale.*

centimetre n: a unit of length in the metric system equal to 0.01 metre (10^{-2} metre). Its symbol is cm.

centipoise n: one-hundredth of a poise.

central facility n: an installation having two or more leases, providing one or more of such functions as separation, compression, dehydration, treating, gathering, or delivery of gas and oil.

centralizer n: also called casing centralizer. See *casing centralizer.*

central oil-treating station n. a processing network used to treat emulsion produced from several leases, thus eliminating the need for individual treating facilities at each lease site.

centrifugal compressor n: a compressor in which the flow of gas to be compressed is moved away from the center rapidly, usually by a series of blades, or turbines. It is a continuous-flow compressor with a low pressure ratio, often used to transmit gas through a pipeline. Gas passing through the compressor contacts a rotating impeller, from which it is discharged into a diffuser, where its velocity is slowed and its kinetic energy changed to static pressure. Centrifugal compressors are nonpositive-displacement machines, often arranged in series on a line to achieve multistage compression.

centrifugal force n: the force that tends to pull all matter from the center of a rotating mass.

centrifugal pump n: a pump with an impeller or rotor, an impeller shaft, and a casing, which discharges fluid by centrifugal force.

centrifuge n: a machine that uses centrifugal force to separate substances of varying densities; also called the shake-out or grind-out machine. A centrifuge is capable of spinning substances at high speeds to obtain high centrifugal forces.

centrifuge test n: a test to determine the amount of BS&W in samples of oil or emulsion. The samples are placed in tubes and spun in a centrifuge, which breaks out the BS&W.

cetane number *n:* a measure of the ignition quality of fuel oil. The higher the cetane number, the more easily the fuel is ignited.

CFG *abbr:* cubic feet of gas; used in drilling reports.

CFR *abbr:* Coordinating Fuels and Equipment Research Committee.

chain *n:* in offshore drilling, a heavy line constructed of iron bars looped together and used for a mooring line.

chain and gear drive *n:* also called chain drive. See *chain drive*.

chain drive *n:* a mechanical drive using a driving chain and chain gears to transmit power. Power transmissions use a roller chain, in which each link is made of side bars, transverse pins, and rollers on the pins. A double roller chain is made of two connected rows of links, a triple roller chain of three, and so forth.

chain tongs *n:* a tool consisting of a handle and releasable chain used for turning pipe or fittings of a diameter larger than that which a pipe wrench would fit. The chain is looped and tightened around the pipe or fitting, and the handle is used to turn the tool so that the pipe or fitting can be tightened or loosened.

change house *n:* a doghouse in which a drilling rig crew changes clothes. See *doghouse*.

change rams *v:* to take rams out of a blowout preventer and replace them with rams of a different size or type. When the size of a drill pipe is changed, the size of the pipe rams must be changed to ensure that they seal around the pipe when closed.

channeling *n:* the bypassing of oil in a water-drive field due to erratic or uncontrolled water encroachment. The natural tendency toward channeling is aggravated by excessive production rates, which encourage premature water encroachment. See *cement channeling*.

charcoal test *n:* a test standardized by the American Gas Association and the Gas Processors Association for determining the natural gasoline content of a given natural gas. The gasoline is adsorbed from the gas on activated charcoal and then recovered by distillation. The test is described in Testing Code 101-43, a joint AGA and GPA publication.

Charles's law *n:* a gas law that states that at constant pressure the volume of a fixed mass or quantity of gas varies directly with the absolute temperature.

charter party *n:* an agreement by which a shipowner agrees to place an entire ship, or part of it, at the disposal of a merchant or other person to carry cargo for an agreed sum.

chase pipe *v:* to lower the drill stem rapidly a few feet into the hole and then stop it suddenly with the drawworks brake. A surge of pressure in the mud in the drill stem and annular space results and may help to flush out debris accumulated in or on the pipe. However, the pressure surge may break down a formation, causing lost circulation, or damage the bit if it is near the bottom.

chase threads *v:* to clean and deburr the threads of a pipe so that it will make up properly.

cheater *n:* a length of pipe fitted over a wrench handle to increase the leverage of the wrench. Use of a large wrench is usually preferred. Also called a snipe.

check valve *n:* a valve that permits flow in one direction only. Commonly referred to as a one-way valve. If the gas or liquid starts to reverse, the valve automatically closes, preventing reverse movement.

Chemelectric treater *n:* a brand name for an electrostatic treater.

chemical cutoff *n:* a method of severing steel pipe in a well by applying high-pressure jets of a very corrosive substance against the wall of the pipe. The resulting cut is very smooth.

chemical pump *n:* an injection pump used to introduce a chemical into a fluid stream or receptacle.

chemical treatment *n:* any of many processes in the oil industry that involve the use of a chemical to effect an operation. Some chemical treatments are acidizing, crude-oil demulsification, corrosion inhibition, paraffin removal, scale removal, drilling fluid control, refinery and plant processes, cleaning and purging operations, waterflood injection, and water purification.

chemisorption *n:* chemical adsorption.

chert *n:* a quartzitic rock with hardness equal to or harder than flint.

chert clause *n:* a provision in a drilling contract stipulating that, when chert is encountered in drilling a well, footage rates are no longer applicable and daywork rates become effective. Chert is very hard and difficult to drill.

chicken hook *n:* a long steel pole with a hook on one end that allows one of the rotary helpers to release the safety latch on the drilling hook so that the bail of the swivel can be removed from the drilling hook (as when the kelly is set back prior to making a trip).

chiller *n:* a heat exchanger that cools process fluids with a refrigerant.

chk *abbr:* choke; used in drilling reports.

chlorine *n:* a greenish yellow, acrid gas (Cl) that causes irritation to the skin and mucous membranes as well as breathing difficulties.

chlorine log *n:* a record of the presence and concentration of chlorine in oil reservoirs, prepared as a method of locating saltwater strata. The log contains both a chlorine and a hydrogen curve. Opposite oil-filled or freshwater zones, the two curves will fall along the same line. Opposite gas-filled zones, the two curves will indicate very low hydrogen density and lower chlorine content.

chlorine survey *n:* a special type of radioactivity-logging survey used inside casing to measure the relative amount of chlorine in the formation outside the casing. Rocks with low chlorine content are likely to contain gas or oil; rocks with high chlorine content usually contain salt water only.

choke *n:* a device with an orifice installed in a line to restrict the flow of fluids. Surface chokes are part of the Christmas tree on a well and contain a choke nipple, or bean, with a small-diameter bore that serves to restrict the flow. Chokes are also used to control the rate of flow of the drilling mud out of the hole when the well is closed in with the blowout preventer and a kick is being circulated out of the hole. See *adjustable choke*, *bottomhole choke*, and *positive choke*.

choke line *n:* an extension of pipe from the blowout preventer assembly, used to direct well fluids from the annulus to the choke manifold.

choke manifold *n:* the arrangement of piping and special valves, called chokes, through which drilling mud is circulated when the blowout preventers are closed and which is used to control the pressures encountered during a kick. See *choke* and *blowout preventer*.

Christmas tree *n:* the control valves, pressure gauges, and chokes assembled at the top of a well to control the flow of oil and gas after the well has been drilled and completed.

chromatograph *n:* an analytical instrument that separates mixtures of substances into identifiable components by means of chromatography.

chromatography *n:* a method of separating a solution of closely related compounds by allowing it to seep through an adsorbent so that each compound becomes adsorbed in a separate layer.

circ *abbr:* circulated; used in drilling reports.

circulate *v:* to pass from one point throughout a system and back to the starting point. For example, drilling fluid is circulated out of the suction pit, down the drill pipe and drill collars, out the bit, up the annulus, and back to the pits while drilling proceeds.

circulate-and-weight method *n:* a method of killing well pressure in which circulation is commenced immediately and mud weight is brought up gradually, according to a definite schedule. Also called concurrent method.

circulating components *n:* the equipment included in the drilling fluid circulating system of a rotary rig. Basically, the components consist of the mud pump, rotary hose, swivel, drill stem, bit, and mud return line.

circulating fluid *n:* also called drilling fluid. See *drilling fluid* and *mud*.

circulating head *n:* an accessory attached to the top of the drill pipe or tubing to form a connection with the mud system to permit circulation of the drilling mud. In some cases, it is also called a rotating head.

circulating pressure *n:* the pressure generated by the mud pumps and exerted on the drill stem.

circulation *n:* the movement of drilling fluid out of the mud pits, down the drill stem, up the annulus, and back to the mud pits.

Cl *sym:* chlorine.

clamp *n:* a mechanical device used to hold an object in place. For example, a leak-repair clamp, or saddle clamp, holds a piece of metal with the same curvature as the pipe over a hole in a line, effecting a temporary seal. A wireline clamp holds the end of a wire rope against the main rope, while a polished-rod clamp attaches the top of the polished rod to the bridle of a pumping unit.

clastic rocks *n pl:* sedimentary rocks composed of fragments of preexisting rocks. Sandstone is a clastic rock.

Claus process *n:* a process to convert hydrogen sulfide into elemental sulfur by selective oxidation.

clay *n:* a fine crystalline material of hydrous silicates, resulting primarily from the decomposition of feldspathic rocks.

clean out *v:* to remove sand, scale, and other deposits from the producing section of the well to restore or increase production.

cleanout door *n:* an opening made to permit removal of sediments from the bottom of a tank. Usually a plate near ground level is removed from the side of the tank to make the door.

cleanout tools *n pl:* the tools or instruments, such as bailers and swabs, used to clean out an oilwell.

clear *v:* to remove brush, trees, rocks, and other obstructions from an area.

clearance volume *n:* the amount of space between the traveling and the standing valves in a sucker rod pump when the pump is at the bottom of its stroke.

clingage *n:* the amount of oil that adheres to the wall of a measuring or prover tank after draining.

Clinton flake *n:* a finely shredded cellophane used as a lost circulation material for cement.

closed circuit *n:* 1. a life-support system in which the gas is recycled continually while the carbon dioxide is removed and oxygen added periodically. 2. a television installation in which the signal is transmitted by wire to a limited number of receivers.

closed-in pressure *n:* See *formation pressure*.

closed system *n:* a water-handling system (such as a saltwater-disposal system) which air is not allowed to enter, used to prevent corrosion or scale.

close in *v:* 1. to temporarily shut in a well that is capable of producing oil or gas. 2. to close the blowout preventers on a well to control a kick. The blowout preventers close off the annulus so that pressure from below cannot flow to the surface.

close nipple *n:* a very short piece of pipe threaded its entire length.

closing machine *n:* a machine that braids wires into strands and strands into rope in the manufacture of wire rope. Also called a stranding machine.

closing ratio *n:* the ratio between the pressure in the hole and the operating-piston pressure needed to close the rams of a blowout preventer.

closing-unit pump *n:* term for an electric or hydraulic pump on an accumulator, serving to pump hydraulic fluid under high pressure to the blowout preventers so that the preventers may be closed or opened.

closure *n:* the vertical distance between the top of an anticline, or dome, and the bottom, an indication of the amount of producing formation that may be expected.

cloud point *n:* the temperature at which paraffin wax begins to congeal and becomes cloudy.

clutch *n:* a coupling used to connect and disconnect a driving and a driven part of a mechanism, especially one that permits the former part to engage the latter gradually and without shock. In the oil field, a clutch permits gradual engaging and disengaging of the equipment driven by a prime mover. *v:* to engage or disengage a clutch.

cm *sym:* centimetre.

cm^2 *sym:* square centimetre.

cm^3 *sym:* cubic centimetre.

CMC *abbr:* carboxymethyl cellulose.

CO *form:* carbon monoxide.

CO$_2$ *form:* carbon dioxide.

coagulation *n:* See *flocculation*.

coal tar epoxy *n:* a thermosetting resin made from the byproduct of the carbonization of bituminous coal, used as a coating because of its adhesiveness, flexibility, and resistance to chemicals.

coating *n:* in corrosion control, any material that forms a continuous film over a metal surface to prevent corrosion damage.

cofferdam *n:* the empty space between two bulkheads separating two adjacent compartments. It is designed to isolate the two compartments from each other, to prevent the liquid contents of one compartment from entering the other in the event of the failure of the bulkhead of one to retain its tightness. In oil tankers, cargo spaces are always isolated from the rest of the ship by cofferdams fitted at both ends of the tank body.

cohesion *n:* the attractive force between the same kinds of molecules (i.e., the force that holds the molecules of a substance together).

coil *n:* an accessory of tubing or pipe for installation in condensers or heat exchangers. In more complex installations, a tube bundle is used instead of a coil.

coke *n:* a solid cellular residue produced from the dry distillation of certain carbonaceous materials, containing carbon as its principal constituent.

coke breeze *n:* crushed coke, used for packing underground anodes in cathodic protection systems to obtain increased anode efficiency at a reduced cost. See *coke*.

cold-work *v:* to work metal without the use of heat.

collapse pressure *n:* the amount of force needed to crush the sides of pipe until it caves in on itself. Collapse occurs when the pressure outside the pipe is greater than the pressure inside the pipe.

collar *n:* 1. a coupling device used to join two lengths of pipe. A combination collar has left-hand threads in one end and right-hand threads in the other. 2. a drill collar. See *drill collar*.

collar locator *n:* a logging device for depth-correlation purposes, operated mechanically or magnetically to produce a log showing the location of each casing collar or coupling in a well. It provides an accurate means of measuring depth in a well.

collar pipe *n:* heavy pipe used between the drill pipe and the bit in the drill stem. See *drill collar*.

collision bulkhead *n:* the foremost bulkhead that extends from the bottom to the freeboard deck of a drill ship. It keeps the main hull watertight if a collision occurs.

colloid *n:* 1. a substance whose particles are so fine that they will not settle out of suspension or solution and cannot be seen under an ordinary microscope. 2. the mixture of a colloid and the liquid, gaseous, or solid medium in which it is dispersed.

colloidal *adj:* pertaining to a colloid; involving particles so minute (less than 2 microns) that they are not visible through optical microscopes. Bentonite is an example of a colloidal clay.

color test *n:* a visual test made against fixed standards to determine the color of a petroleum or other product.

column-stabilized, semisubmersible drilling rig *n:* a semisubmersible drilling rig that has a large lower hull, boxlike or tubular in shape, with several watertight columns extending to an upper deck on which the drilling machinery is located. Such a rig can drill with the lower hull resting on bottom or submerged below the water and floating. Usually the distance between the lower hull and the upper deck is fixed.

combination drive *n:* the natural energy that forces fluids from a reservoir and into a wellbore, provided by a gas cap above and water below the oil in the reservoir. See *reservoir drive mechanism*, *gas-cap drive*, and *water drive*.

combination rig *n:* a light rig that has the essential elements for both rotary and cable-tool drilling. It is sometimes used for reconditioning wells.

combination string *n:* a casing string with joints of various collapse resistance, internal yield strength, and tensile strength, designed for various depths in a specific well to best withstand the conditions of that well. In deep wells, high tensile strength is required in the top casing joints to carry the load, whereas high collapse resistance and internal yield strength are needed in the bottom joints. In the middle of the casing, average qualities are usually sufficient. The most suitable combination of types and weights of pipe helps to ensure efficient production at a minimum cost.

combination trap *n:* a subsurface hydrocarbon trap that has the features of both a structural trap and a stratigraphic trap.

combustion *n:* 1. the process of burning. Chemically, it is a process of rapid oxidation caused by the union of oxygen from the air with the material that is being oxidized or burned. 2. the organized and orderly burning of fuel inside the cylinder of an engine.

come-along *n:* a manually operated device that is used to tighten guy wires or move heavy loads. Usually, a come-along is a gripping tool with two jaws attached to a ring so that when the ring is pulled, the jaws close.

come in *v:* to begin to produce; to become profitable.

come out of the hole *v:* to pull the drill stem out of the wellbore. This withdrawal is necessary to change the bit, change from a core barrel to the bit, run electric logs, prepare for a drill stem test, run casing, and so on.

come to see you *v:* (slang) to blow out; to kick. A well will "come to see you" if it blows out.

commercial butane *n:* a liquefied hydrocarbon consisting chiefly of butane or butylenes and conforming to the GPA specification for commercial butane defined in GPA Publication 2140.

commercial production *n:* oil and gas production of sufficient quantity to justify keeping a well in production.

commercial propane *n:* a liquefied hydrocarbon product consisting chiefly of propane and/or propylene and conforming to the GPA specification for commercial propane as defined in GPA Publication 2140.

commercial quantity *n:* an amount of oil and gas large enough to justify the expense of producing it.

commingling *n:* the mixing together of crude oil products that have similar properties, usually for convenient transportation in a pipeline.

common carrier *n:* any cargo transportation system available for public use. Nearly all pipelines are common carriers.

common cement *n:* a regular portland cement classified as either API Class A or ASTM Type 1 cement.

common rail *n:* the line in a certain type of fuel-injection system for a diesel engine that keeps fuel at a given pressure and feeds it through feed lines to each fuel injector.

common-rail injection *n:* a fuel-injection system on a diesel engine in which one line, or rail, holds fuel at a certain pressure, and feed lines run from it to each fuel injector.

commutator *n:* a series of bars connected to the armature coils of an electric motor or generator. As the commutator rotates in contact with fixed brushes, the direction of flow of current to or from the armature is in one direction only.

comp *abbr:* completed or completion; used in drilling reports.

compact *n:* also called an insert. See *insert*.

company man *n:* also called company representative. See *company representative*.

company representative *n:* an employee of an operating company whose job is to represent the company's interests at the drilling location.

compartment *n:* a subdivision of space on a floating offshore drilling rig, a ship, or a barge.

compensated neutron log *n:* a neutron log in which there are a source and two detectors. The ratio of the count rates from the two detectors is processed by a small computer which calculates apparent limestone porosity. The compensated neutron log has fewer borehole effects than any other type of neutron log.

complete a well *v:* to finish work on a well and bring it to productive status. See *well completion*.

completion fluid *n:* a special drilling mud used when a well is being completed. It is selected not only for its ability to control formation pressure, but also for the properties that minimize formation damage.

composite sample *n:* a sample of a substance that is a mixture or solution of several other substances. In a crude oil storage tank, a composite sample is taken at the top, at the bottom, and in the middle.

composite stream *n:* a flow of oil and gas in one stream; a flow of two or more different liquid hydrocarbons in one stream.

compound *n:* 1. a mechanism used to transmit power from the engines to the pump, the drawworks, and other machinery on a drilling rig. It is composed of clutches, chains and sprockets, belts and pulleys, and a number of shafts, both driven and driving. 2. a substance formed by the chemical union of two or more elements in definite proportions; the smallest particle of a chemical compound is a molecule. *v:* to connect two or more power-producing devices, such as engines, to run driven equipment, such as the drawworks.

compressibility factor *n:* a factor, usually expressed as Z, which gives the ratio of the actual volume of gas at a given temperature and pressure to the volume of gas when calculated by the ideal gas law without any consideration of the compressibility factor.

compression *n:* the act or process of squeezing a given volume of gas into a smaller space.

compression ignition *n:* an ignition method used in diesel engines by which the air in the cylinder is compressed to such a degree by the piston that ignition occurs upon the injection of fuel. About a 1-lb (7 kPa) rise in pressure causes a 2°F (1°C) increase in temperature.

compression pressure *n:* the pounds-per-square-inch (kilopascal) increase in pressure at the end of the compression stroke in an engine, about 500 psi (3 500 kPa) in a diesel engine.

compression ratio *n:* 1. the ratio of the absolute discharge pressure from a compressor to the absolute intake pressure. 2. the ratio of the volume of an engine cylinder before compression to its volume after compression. For example, if a cylinder volume of 10 in.^{3} (10 cm^3) is compressed into 1 in.^{3} (1 cm^3), the compression ratio is 10:1.

compression-refrigeration cycle *n:* the refrigeration cycle in which refrigeration is supplied by the evaporation of a liquid refrigerant such as propane or ammonia. Compare *absorption-refrigeration cycle*.

compressive strength *n:* the degree of resistance of a material to a force acting along one of its axes in a manner tending to collapse it; usually expressed in pounds of force per square inch (psi) of surface affected or in kilopascals.

compressor n: a device that raises the pressure of a compressible fluid such as air or gas. Compressors create a pressure differential to move or compress a vapor or a gas, consuming power in the process. They may be positive-displacement compressors or nonpositive-displacement compressors. See *centrifugal compressor, reciprocating compressor,* and *jet compressor.*

compressor clearance n: the ratio of the volume remaining in a compressor cylinder at the end of a compression stroke to the volume displaced by one stroke of the piston. The ratio is usually expressed in percent.

compressor station n: a facility consisting of one or more compressors with the necessary auxiliaries for delivering compressed gas.

Compton effect n: a reaction in which gamma rays with intermediate energy levels (0.6 to 2.5 mev) lose their energy by collision with orbital electrons.

computer n: a machine capable of processing information or providing data by automatically following preprogrammed directions.

computer program n: a set of instructions fed into a computer to perform a task or to solve a problem.

concession n: a tract of land granted by a government to an individual or a company for exploration and exploitation in recovering minerals.

concurrent method n: also called circulate-and-weight method. See *circulate-and-weight method.*

condensate n: a light hydrocarbon liquid obtained by condensation of hydrocarbon vapors. It consists of varying proportions of butane, propane, pentane, and heavier fractions, with little or no methane or ethane.

condensate reservoir n: a reservoir in which both condensate and gas exist in one homogeneous phase. When fluid is drawn from such a reservoir and the pressure decreases below the critical level, a liquid phase (condensate) appears.

condensate well n: a gas well producing from a condensate gas reservoir.

condensation n: the process by which vapors are converted into liquids, chiefly accomplished by cooling the vapors. Condensation is often the cause of water appearing in fuels.

condenser n: 1. a form of heat exchanger in which the heat in vapors is transferred to a flow of cooling water or air, causing the vapors to form a liquid. 2. a capacitor.

condition v: to treat drilling mud with additives to give it certain properties. Sometimes the term applies to water used in boilers, drilling operations, and so on. To condition and circulate mud is to ensure that additives are distributed evenly throughout a system by circulating the mud while it is being conditioned.

conductivity n: 1. the ability to transmit or convey (as heat or electricity). 2. an electrical logging measurement obtained from an induction survey, in which eddy currents produced by an alternating magnetic field induce in a receiver coil a voltage proportionate to the ability of the formation to conduct electricity.

conductor line n: a small-diameter conductive line used in electric wireline operations, such as electric well logging and perforating, in which the transmission of electrical current is required. Compare *wireline.*

conductor pipe n: 1. a short string of large-diameter casing used to keep the wellbore open and to provide a means of conveying the upflowing drilling fluid from the wellbore to the mud pit. 2. a boot. See *boot.*

cone n: a conical-shaped metal device into which cutting teeth are formed or mounted on a roller cone bit. See *roller cone bit.*

cone bit n: a roller bit in which the cutters are conical. See *bit.*

cone offset n: the amount by which lines drawn through the center of each cone of the bit fail to meet in the center of the bit. For example, in a roller cone bit with three cones, three lines can be drawn through the center of each cone and extended to the center of the bit. If these cone centerlines do not meet in the bit's center, the cones are said to be offset.

cone-roof tank n: a tank with a fixed conical roof.

cone shake n: shaking or vibrating of the cones of a bit.

cone shell n: that part of the cone of a roller cone bit out of which the teeth are milled or into which tungsten carbide inserts are inserted and inside of which are housed the bearings.

cone skidding n: locking of a cone on a roller cone bit so that it will not turn when the bit is rotating. Cone skidding results in a flattening of the surface of the cone in contact with the bottom of the hole.

confirmation well *n:* the second producer in a new field, following the discovery well.

conformable *adj:* layered in parallel and unbroken rows of rock, indicating that no disturbance occurred during deposition of the rock. Compare *unconformity*.

congl *abbr:* conglomerate; used in drilling reports.

conglomerate *n:* a sedimentary rock composed of pebbles of various sizes held together by a cementing material such as clay. Conglomerates are similar to sandstone but have larger grains.

conical angle *n:* the angle of the cone of a bit. This angle may be steep, in which case the cone has a sharp taper, or it may be shallow, in which case the cone has a flatter taper.

coning *n:* the encroachment of reservoir water into the oil column and well because of uncontrolled production.

connate water *n:* water retained in the pore spaces, or interstices, of a formation from the time the formation was created. Compare *interstitial water*.

connecting rod *n:* 1. a forged-metal shaft that joins the piston of an engine to the crankshaft. 2. the metal shaft that is joined to the bull gear and crosshead of a mud pump.

connecting rod bearing *n:* the bearing between the rod and the crankshaft; often called the rod bearing.

connection *n:* 1. a section of pipe or fitting used to join pipe to pipe or pipe to a vessel. 2. a place in electrical circuits where wires join together.

connection gas *n:* the relatively small amount of gas that enters a well when the mud pump is stopped in order that a connection may be made.

conservation *n:* preservation; economy; avoidance of waste. It is especially important in the petroleum industry, since oil and gas are irreplaceable. Many conservation practices, such as the trapping of condensable vapors, are used in the industry.

consistency *n:* the cohesion of the individual particles of a given material (i.e., its ability to deform or its resistance to flow).

console *n:* See *driller's console*.

constant choke-pressure method *n:* a method of killing a well that has kicked, in which the choke size is adjusted to maintain a constant casing pressure. This method does not work unless the kick is all or nearly all salt water; if the kick is gas, this method will not maintain a constant bottomhole pressure because gas expands as it rises in the annulus.

constant pit-level method *n:* a method of killing a well in which the mud level in the pits is held constant while the choke size is reduced and the pump speed slowed. It is not effective because casing pressure increases to the point where the formation fractures or casing ruptures, and control of the well is lost.

contactor *n:* a vessel or piece of equipment in which two or more substances are brought together.

contaminant *n:* a material, usually a mud component, that becomes mixed with cement slurry during displacement and affects it adversely.

continental margin *n:* a zone that separates emergent continents from the deep sea bottom.

continental shelf *n:* a zone, adjacent to a continent, that extends from the low waterline to the point at which the seafloor slopes off steeply to 600 feet (183 m) deep or more.

continuous-flow gas lift *n:* See *gas lift*.

continuous flowmeter log *n:* a log used to determine the contribution of each zone to the total production or injection. These surveys are used to indicate changes in the flow pattern versus changes in conditions at the surface, in time, in type of operation, or after stimulation treatments; particularly useful for measuring gas well flow.

continuous phase *n:* the liquid in which solids are suspended or droplets of another liquid are dispersed; sometimes called the external phase. In a water-in-oil emulsion, oil is the continuous phase. Compare *internal phase*.

continuous treatment *n:* a method of applying corrosion inhibitors to production fluids, in which the concentration of the inhibitor can be maintained at constant levels.

contour map *n:* a map that has lines marked to indicate points or areas that are the same elevation above or below sea level. It is often used by geologists to depict subsurface features.

contract *n:* an agreement, usually written, listing the terms under which services are to be performed. A drilling contract covers such factors as the cost of drilling the well (whether by foot or by day), the distribution of expenses between operator and contractor, and the type of equipment to be used.

contract depth *n:* the depth of the wellbore at which a drilling contract is fulfilled.

control board *n:* a panel on which are grouped various control devices such as switches and levers, along with indicating instruments.

control pod *n:* See *hydraulic control pod*.

control valve *n:* a valve designed to regulate the flow or pressure of a fluid.

conventional completion *n:* a method for completing a well in which tubing is set inside 4½-inch or larger casing. Compare *miniaturized completion*.

conventional gas-lift mandrel *n:* See *gas-lift mandrel*.

convergence pressure *n:* the pressure at a given temperature for a hydrocarbon system of fixed composition at which the vapor-liquid equilibria values of the various components in the system become or tend to become unity. The convergence pressure is used to adjust vapor-liquid equilibria values to the particular system under consideration.

conversion *n:* the change in the chemistry of a mud from one type to another; also called a breakover. Reasons for making a conversion may be (1) to maintain a stable wellbore, (2) to provide a mud that will tolerate higher weight, or density, (3) to drill soluble formations, and (4) to provide protection to producing zones.

coolant *n:* a cooling agent, usually a fluid, such as the liquid applied to the edge of a cutting tool to carry off frictional heat or a circulating fluid for cooling an engine.

cooler *n:* a heat exchanger that reduces the temperature of a fluid by transferring the heat to a nonprocess medium.

cooling tower *n:* a structure in which air contact is used to cool a stream of water that has been heated by circulation through a system. The air flows countercurrently or crosscurrently to the water.

cooling water *n:* treated fresh water that circulates inside a diesel engine to transfer heat.

Coordinating Fuels and Equipment Research Committee *n:* a committee composed of engine manufacturing, petroleum refining, petroleum consuming, university, government, and other technical persons who supervise cooperative testing and study of engine fuels for the Coordinating Research Council, Inc.

Coordinating Research Council, Inc. *n:* a nonprofit organization supported jointly by the American Petroleum Institute and the Society of Automotive Engineers, Inc. It administers work of the CRF and other committees that correlate test work and other studies on fuels, lubricants, engines, and engine equipment.

copper strip test *n:* a test using a small strip of pure copper to determine qualitatively the corrosivity of a product.

copper sulfate electrode *n:* a commonly used nonpolarizing electrode used in corrosion control to measure the electrical potential of a metal structure to a surrounding electrolyte in order to determine the potential for corrosion damage or to monitor the effectiveness of existing control measures. See *half-cell*.

cordage *n:* all of the rope on a ship or an offshore drilling rig.

core *n:* a cylindrical sample taken from a formation for geological analysis. Usually a conventional core barrel is substituted for the bit and procures a sample as it penetrates the formation. *v:* to obtain a formation sample for analysis.

core analysis *n:* laboratory analysis of a core sample to determine porosity, permeability, lithology, fluid content, angle of dip, geological age, and probable productivity of the formation.

core barrel *n:* a tubular device, usually from 10 to 60 feet long, run at the bottom of the drill pipe in place of a bit and used to cut a core sample.

core catcher *n:* the part of the core barrel that holds the formation sample.

core cutterhead *n:* the cutting element of the core barrel assembly. In design it corresponds to one of the three main types of bits: drag bits with blades for cutting soft formations; roller bits with rotating cutters for cutting medium-hard formations; and diamond bits for cutting very hard formations.

core-drill *v:* to drill shallow, small-diameter wells to obtain geological information, usually in the bottom of an existing wellbore. A continuous sample of the formation is provided from the top to the final depth.

coring reel *n:* also called a sand reel. See *sand reel*.

Corod *n:* a trade name for a special form of sucker rod. Corod, or continuous rod, normally has no joints between the downhole pump and the surface. See *sucker rod*.

correlate *v:* to relate subsurface information obtained from one well to that of others so that the formations may be charted and their depths and thicknesses noted. Correlations are made by comparing electrical well logs, radioactivity logs, and cores from different wells.

corrosion *n:* any of a variety of complex chemical or electrochemical processes by which metal is destroyed through reaction with its environment. For example, rust is corrosion. See *corrosion cell*.

corrosion cell *n:* the pattern of flow of electric current between metals in electrolyte, which causes metal to corrode or deteriorate.

corrosion control *n:* the measures used to prevent or reduce the effects of corrosion. These practices can range from simply painting metal, to isolate it from moisture and chemicals and to insulate it from galvanic currents, to cathodic protection, in which a galvanic or impressed direct electric current renders a pipeline cathodic, thus causing it to be a negative element in the circuit. The use of chemical inhibitors and closed systems are other examples of corrosion control.

corrosion coupon *n:* a metal strip inserted into a system to monitor corrosion rate and to indicate corrosion-inhibitor effectiveness.

corrosion fatigue *n:* metal fatigue concentrated in corrosion pits. See *fatigue*.

corrosion test *n:* one of a number of tests to determine qualitatively or quantitatively the corrosion-inducing compounds in a product.

corrosive product *n:* a hydrocarbon product which contains corrosion-inducing compounds in excess of the specification limits for a sweet product.

coulomb *n:* the metric unit of electric charge, having the symbol C.

counterbalance system *n:* also called the two-step grooving system. See *two-step grooving system*.

counterbalance weight *n:* a weight applied to compensate for existing weight or force. On pumping units in oil production, counterweights are used to offset the weight of the column of sucker rods and oil on the upstroke of the pump, and the weight of the rods on the downstroke.

counterbore *n:* flat-bottomed enlargement of the mouth of a cylindrical bore. *v:* to enlarge part of a hole by means of a counterbore.

countercurrent stripping *n:* the use of natural or inert gas to remove oxygen from production systems.

countershaft *n:* a shaft that gets its movement from a main shaft and transmits it to a working part.

coupling *n:* 1. in piping, a metal collar with internal threads used to join two sections of threaded pipe. 2. in power transmission, a connection extending longitudinally between a driving shaft and a driven shaft. Most such couplings are flexible and compensate for minor misalignment of the two shafts.

coupon *n:* See *corrosion coupon*.

CP *abbr:* casing pressure or casing point; used in drilling reports.

cp *sym:* centipoise.

CPC *abbr:* computerized production control.

crack a valve *v:* to open a valve so that it leaks just a little.

cracking *n:* in petroleum refining, the process of breaking down large chemical compounds into smaller compounds. The two major types of cracking are thermal cracking and catalytic cracking. See *thermal cracking* and *catalytic cracking*.

crane *n:* a machine for raising, lowering, and revolving heavy pieces of equipment, especially on offshore rigs and platforms.

crankcase *n:* the housing that encloses the crankshaft of an engine.

crankshaft *n:* a rotating shaft to which connecting rods are attached. It changes up and down (reciprocating) motion into circular (rotary) motion.

crater *v:* (slang) to cave in; to fail. After a violent blowout, the force of the fluids escaping from the wellbore sometimes blows a large hole in the ground. In this case, the well is said to have cratered. Equipment craters when it fails.

CRC *abbr:* Coordinating Research Council, Inc.

crd *abbr:* cored; used in drilling reports.

crew *n:* the workers on a drilling or workover rig, including the driller, derrickman, and rotary helpers.

crew chief *n:* the driller or head well pusher in charge of operations on a well servicing rig employed to pull sucker rods or tubing.

crg *abbr:* coring; used in drilling reports.

critical density *n:* the density of a substance at the critical temperature and pressure.

critical point *n:* 1. the point at which, in terms of temperature and pressure, a fluid cannot be distinguished as being either a gas or a liquid; the point at which the physical properties of a liquid and a gas are identical. 2. one of the places along the length of drilling line at which strain is exerted as pipe is run into or pulled out of the hole.

critical pressure *n:* the pressure needed to condense a vapor at its critical temperature.

critical speed n: the speed reached by an engine or rotating system that corresponds to a resonance frequency of the engine or system. Often, in combination with power impulses, critical speed can cause damaging shock waves.

critical temperature n: the highest temperature at which a substance can be separated into two fluid phases—liquid and vapor. Above the critical temperature, a gas cannot be liquefied by pressure alone.

critical weight n: the weight placed on the bit that results in a tension on the drill string causing the drill string to become resonant at the rotary speed being used. A drill stem operating with critical weight and at the critical speed for that weight will have stresses developed that cause very rapid failure.

crooked hole n: a wellbore that has deviated from the vertical. It usually occurs where there is a section of alternating hard and soft strata steeply inclined from the horizontal.

crooked-hole country n: a geographical area in which the subsurface formations are so arranged that it is difficult to drill a hole straight through them. See *crooked hole*.

crosshead n: the block in a mud pump that is guided to move in a straight line and serves as a connection between the pony rod and the connecting rod.

crossover n: the section of a drawworks drum grooved for angle control in which the wire rope crosses over to start a new wrap. Also called an angle-control section.

crossover joint n: a length of casing with one thread on the field end and a different thread in the coupling, used to make a changeover from one thread to another in a string of casing.

crossover sub n: a sub used between two sizes or types of threads in the drill stem assembly.

cross section n: the property of atomic nuclei of having the probability of collision with a neutron. The nucleus of a lighter element is more likely to collide with a neutron than the nucleus of a heavier element. Cross section varies with the elements and with the energy of the neutron.

cross-thread v: to screw together two threaded pieces when the threads of the pieces have not been aligned properly.

crown n: 1. the crown block or top of a derrick or mast. See *crown block*. 2. the top of a piston. See *piston*. 3. a high spot formed on a tool joint shoulder as the result of wobble.

crown block n: an assembly of sheaves, mounted on beams at the top of the derrick, over which the drilling line is reeved. See *block*.

crown frame flanges n: projections on the frame to which the crown block is attached.

Crown-O-Matic n: a brand name for a special air-relay valve mounted near the crown that, when struck by the traveling block, conveys air pressure to the air brakes of the drawworks to prevent the traveling block from striking the crown.

crown platform n: the working platform at the top of the derrick that permits access to the sheaves of the crown block.

crow's nest n: an elevated walkway where employees work (as on the top of a derrick or a refinery tower).

crude oil n: unrefined liquid petroleum. It ranges in gravity from 9° API to 55° API and in color from yellow to black, and it may have a paraffin, asphalt, or mixed base. If a crude oil, or crude, contains a sizable amount of sulfur or sulfur compounds, it is called a sour crude; if it has little or no sulfur, it is called a sweet crude. In addition, crude oils may be referred to as heavy or light according to API gravity, the lighter oils having the higher gravities.

cryogenic plant n: a gas processing plant that is capable of producing natural gas liquid products, including ethane, at very low operating temperatures.

cryogenics n: the study of the effects of very low temperatures.

cu abbr: cubic.

cubic centimetre n: a commonly used unit of volume measurement in the metric system equal to 10^{-6} cubic metre, or 1 millilitre. The symbol for cubic centimetre is cm^3.

cubic foot n: the volume of a cube, all edges of which measure 1 foot. Natural gas in the United States is usually measured in cubic feet, with the most common standard cubic foot being measured at 60°F and 14.65 psia, although base conditions vary from state to state.

cubic metre n: a unit of volume measurement in the metric system, replacing the previous standard unit known as the barrel, which was equivalent to 35 imperial gallons or 42 United States gallons. The cubic metre equals approximately 6.2898 barrels.

cup n: a low spot formed on a tool joint shoulder as the result of wobble.

cup packer n: a device made up in the drill stem, lowered into the well in order to allow the casing has four legs standing at the corners of the substructure and reaching to the crown block. The substructure is an assembly of heavy beams used to elevate the derrick and provide space to install blowout preventers, casingheads, and so forth. Because the standard derrick must be assembled piece by piece, it has largely been replaced by the mast, which can be lowered and raised without disassembly.

derrick floor *n:* also called the rig floor or the drill floor. See *rig floor*.

derrickman *n:* the crew member who handles the upper end of the drill string as it is being hoisted out of or lowered into the hole. He is also responsible for the circulating machinery and the conditioning of the drilling fluid.

desalt *v:* to remove dissolved salt from crude oil. Sometimes fresh water is injected into the crude stream to dissolve salt for removal by electrostatic treaters.

desander *n:* a centrifugal device for removing sand from drilling fluid to prevent abrasion of the pumps. It may be operated mechanically or by a fast-moving stream of fluid inside a special cone-shaped vessel, in which case it is sometimes called a hydrocyclone. Compare *desilter*.

desiccant *n:* a substance able to remove water from another substance with which it is in contact. It may be liquid (as triethylene glycol) or solid (as silica gel).

design factor *n:* the ratio of the ultimate load a vessel or structure will sustain to the permissibly safe load placed on it. Such safety factors are incorporated into the design of casing, for example, to allow for unusual burst, tension, or collapse stresses.

design factor of wire rope *n:* also called safety factor. See *safety factor of wire rope*.

design water depth *n:* 1. the vertical distance from the ocean bottom to the nominal water level plus the height of astronomical and storm tides. 2. the deepest water in which an offshore drilling rig can operate.

desilter *n:* a centrifugal device for removing very fine particles, or silt, from drilling fluid to keep the amounts of solids in the fluid at the lowest possible point. Usually, the lower the solids content of mud, the faster is the rate of penetration. The desilter works on the same principle as a desander. Compare *desander*.

Desk and Derrick Club *n:* an association of women employed in the petroleum and allied industries. The principal function of the group is that of providing informational and educational programs for the enlightenment of its members about the industry they serve. Membership ranges from secretaries through managers and directors of companies.

desulfurize *v:* to remove sulfur or sulfur compounds from oil or gas.

detonation *n:* 1. an explosion. 2. the knock or ping produced when fuel of too-low octane rating is used in the engine. Compare *preignition*.

deuterium *n:* the isotope of the element hydrogen that has one neutron and one proton in the nucleus; atomic weight is 2.0144.

development well *n:* 1. a well drilled in proven territory in a field to complete a pattern of production. 2. an exploitation well. See *exploitation well*.

deviation *n:* the inclination of the wellbore from the vertical. The angle of deviation, angle of drift, or drift angle is the angle in degrees that shows the variation from the vertical as revealed by a deviation survey. See *deviation survey*.

deviation survey *n:* an operation made to determine the angle from which a bit has deviated from the vertical during drilling. There are two basic deviation-survey, or drift-survey, instruments: one reveals the angle of deviation only; the other indicates both the angle and the direction of deviation.

dew point *n:* the temperature and pressure at which a liquid begins to condense out of a gas. For example, if a constant pressure is held on a certain volume of gas but the temperature is reduced, a point is reached at which droplets of liquid condense out of the gas. That point is the dew point of the gas at that pressure. Similarly, if a constant temperature is maintained on a volume of gas but the pressure is increased, the point at which liquid begins to condense out is the dew point at that temperature. Compare *bubble point*.

dew-point recorder *n:* a device used by gas transmission companies to determine and to record continuously the dew point of the gas.

DF *abbr:* derrick floor; used in drilling reports.

diagenesis *n:* the chemical and physical changes that sediments undergo (after deposition, compaction, cementation, recrystallization, and sometimes replacement) that result in lithification.

diameter *n:* the distance across a circle, measured through its center. In the measurement of pipe diameters, the inside diameter is that of the interior

diamond bit-direct connection

circle and the outside diameter that of the exterior circle.

diamond bit n: a drilling bit that has a steel body surfaced with a matrix and industrial diamonds. Cutting is performed by the rotation of the very hard diamonds over the rock surface.

diatomaceous earth n: an earthy deposit made up of the siliceous cell walls of one-celled marine algae called diatoms, used as an admixture for cement to produce a low-density slurry.

die n: a tool used to shape, form, or finish other tools or pieces of metal. For example, a threading die is used to cut threads on pipe.

die collar n: a collar or coupling of tool steel, threaded internally, that is used to retrieve pipe from the well on fishing jobs; the female counterpart of a taper tap. The die collar is made up on the drill pipe and lowered into the hole until it contacts the lost pipe. Rotation of the die collar on top of the pipe cuts threads on the outside of the pipe, providing for a firm attachment. The pipe is then retrieved from the hole. Compare *taper tap*.

dielectric n: a substance that is an insulator, or nonconductor, of electricity.

dielectric constant n: the value of dielectricity assigned to a substance. A substance that is a good insulator has a high dielectric constant, while a poor insulator has a low one. The dielectric constant of oil is lower than that of water, and upon this principle a net-oil computer operates.

die nipple n: a device similar to a die collar but with external threads.

diesel-electric drilling rig n: a drilling rig that is powered by diesel engines driving electric generators. Commonly used offshore, and becoming increasingly used on land.

diesel-electric power n: the power supplied to a drilling rig by diesel engines driving electric generators, used widely offshore and gaining popularity onshore.

diesel engine n: a high-compression, internal-combustion engine used extensively for powering drilling rigs. In a diesel engine, air is drawn into the cylinders and compressed to very high pressures; ignition occurs as fuel is injected into the compressed and heated air. Combustion takes place within the cylinder above the piston, and expansion of the combustion products imparts power to the piston.

diesel fuel n: a light hydrocarbon mixture for diesel engines, similar to furnace fuel oil; it has a boiling range just above that of kerosine.

differential n: the difference in quantity or degree between two measurements or units. For example, the pressure differential across a choke is the variation between the pressure on one side and that on the other.

differential fill-up collar n: a device used in setting casing. It is run near the bottom of the casing to admit drilling fluids automatically into the casing as needed; it sinks rather than floats in the well.

differential pressure n: the difference between two fluid pressures; for example, the difference between the pressure in a reservoir and in a wellbore drilled in the reservoir, or between atmospheric pressure at sea level and at 10,000 feet. Also called pressure differential.

differential-pressure gauge n: a pressure-measuring device actuated by two or more pressure-sensitive elements that act in opposition to produce an indication of the difference between two pressures.

differential-pressure sticking n: a condition in which the drill stem becomes stuck against the wall of the wellbore because part of the drill stem (usually the drill collars) has become embedded in the filter cake. Necessary conditions for differential-pressure sticking, or wall sticking, are a permeable formation and a pressure differential across a nearly impermeable filter cake and the drill stem. See *differential pressure*.

diffusion n: 1. the spontaneous movement and intermingling of particles of liquids, gases, or solids. 2. the migration of dissolved substances to areas of least concentration.

dip n: also called formation dip. See *formation dip*.

dip log n: a survey of a wellbore with a dipmeter, made to determine the direction and angle of dip of certain formations exposed to the wellbore.

dipmeter survey n: an oilwell-surveying method that determines the direction and angle of formation dip in relation to the borehole. It records data that permit computation of both the amount and direction of formation dip relative to the axis of the hole and thus provides information about the geologic structure of the formation.

direct connection n: a straightforward connection that makes the speeds of a prime mover and a driven machine identical.

direct current *n:* electric current that flows in only one direction.

directional drilling *n:* intentional deviation of a wellbore from the vertical. Although wellbores are normally drilled vertically, it is sometimes necessary or advantageous to drill at an angle from the vertical. Controlled directional drilling makes it possible to reach subsurface areas laterally remote from the point where the bit enters the earth. It often involves the use of turbodrills, Dyna-Drills, whipstocks, or other deflecting tools.

directional engineer *n:* an engineering specialist in directional drilling, usually employed by a service company specializing in directional drilling.

directional hole *n:* a wellbore intentionally drilled at an angle from the vertical. See *directional drilling*.

directional operator *n:* a practical or trained technician who supervises rig operations relating to directional drilling, usually employed by a company specializing in this service.

directional survey *n:* a logging method that records hole drift, or deviation from the vertical, and direction of the drift. A single-shot directional-survey instrument makes a single photograph of a compass reading of the drift direction and the number of degrees the hole is off vertical. A multishot-survey instrument obtains numerous readings in the hole as the device is pulled out of the well. See *directional drilling*.

discharge line *n:* a line through which drilling mud travels from the mud pump to the standpipe on its way to the wellbore.

disconformity *n:* an unconformity in which formations on opposite sides of it are parallel.

discovery well *n:* the first oil or gas well drilled in a new field; the well that reveals the presence of a petroleum-bearing reservoir. Subsequent wells are development wells.

dispatcher *n:* an employee responsible for scheduling movement of oil through pipelines.

dispersant *n:* a substance added to cement that chemically wets the cement particles in the slurry, allowing the slurry to flow easily without much water.

dispersible inhibitor *n:* an inhibitor substance that can be evenly dispersed in another liquid with moderate agitation.

dispersion *n:* a suspension of extremely fine particles in a liquid (such as colloids in a colloidal solution).

displacement *n:* the weight of a fluid (such as water) displaced by a freely floating or submerged body (such as an offshore drilling rig). If the body floats, the displacement equals the weight of the body.

displacement fluid *n:* in oilwell cementing, the fluid, usually drilling mud or salt water, that is pumped into the well after the cement is pumped into it to force the cement out of the casing and into the annulus.

displacement meter *n:* a meter in which a piston is actuated by the pressure of a measured volume of liquid, and the volume swept by the piston is equal to the volume of the liquid recorded.

displacement plunger *n:* a device used to pump liquids, usually at high pressures, with an action similar to that of a piston.

displacement rate *n:* a measurement of the speed with which a volume of cement slurry or mud is pumped down the hole.

disposal well *n:* a well into which salt water or spent chemical is pumped, most commonly part of a saltwater-disposal system.

dissociation *n:* the separation of a molecule into two or more fragments (atoms, ions) by interaction with another body or by the absorption of electromagnetic radiation.

dissolved-gas drive *n:* a solution-gas drive. See *reservoir drive mechanism*.

distillate *n:* 1. a product of distillation; the liquid condensed from the vapor produced in a still. Sometimes called condensate. See *condensate*. 2. heavy gasoline or light kerosines used as fuels.

distillation *n:* the process of driving off gas or vapor from liquids or solids, usually by heating, and condensing the vapor back to liquid to purify, fractionate, or form new products.

distribution *n:* the apportioning of daily production rates to wells on a lease. Because there are many wells on a lease, such production is apportioned on the basis of periodic tests rather than on the individual receiving and gauging of oil at each well.

distribution system *n:* a system of pipelines and other equipment by which natural gas or other products are distributed to customers, to lease operations, or to other points of consumption.

distributor *n:* a device that directs the proper flow of fuel to the proper place at the proper time in the proper amount.

disulfides *n pl:* chemical compounds containing an -S-S- linkage. They are colorless liquids completely miscible with hydrocarbons, insoluble in water, and sweet to the doctor test. Mercaptans are converted to disulfides in treating processes employing oxidation reactions.

ditch *n:* a trench or channel made in the earth, usually to bury pipeline, cable, etc. On a drilling rig, the mud flow channel from the conductor-pipe outlet is often called a ditch. See *mud return line.*

diverter *n:* a system used to control well blowouts encountered at relatively shallow depths and to protect floating rigs during blowouts by directing the flow away from the rig.

diving bell *n:* a cylindrical or spherical compartment used to transport a diver or dive team to and from an underwater work site.

dk *abbr:* dark; used in drilling reports.

doctor test *n:* a qualitative method for detecting hydrogen sulfide and mercaptans in petroleum distillates. The test distinguishes between sour and sweet products.

doghouse *n:* 1. a small enclosure on the rig floor, used as an office for the driller or as a storehouse for small objects. 2. any small building used as an office, a change house, or a place for storage.

dogleg *n:* 1. a short change of direction in the wellbore, frequently resulting in the formation of a key seat. See *key seat.* 2. a sharp bend permanently put in an object such as a pipe.

dol *abbr:* dolomite; used in drilling reports.

dolly *n:* See *pipe dolly.*

dolo *abbr:* dolomite; used in drilling reports.

dolomite *n:* a type of sedimentary rock similar to limestone but rich in magnesium carbonate; sometimes a reservoir rock for petroleum.

dolomitization *n:* the shrinking of the solid volume of rock as limestone turns to dolomite; the conversion of limestone to dolomite rock by replacement of a portion of the calcium carbonate with magnesium carbonate.

dome *n:* a geologic structure resembling an inverted bowl; a short anticline that plunges on all sides.

dome plug trap *n:* a reservoir formation in which fluid or plastic masses of rock material originated at unknown depths and pierced or lifted the overlying sedimentary strata.

dome-roof tank *n:* a storage tank with a dome-shaped roof affixed to the shell.

doodlebug *n:* (slang) 1. the seismograph used in prospecting for potential oil-bearing geological structures. 2. any of the various devices used in searching for petroleum deposits.

door sheet *n:* a plate at the base of a tank shell or wall that is removed to allow the tank to be cleaned.

dope *n:* a lubricant for the threads of oil field tubular goods.

double *n:* a length of drill pipe, casing, or tubing consisting of two joints screwed together. Compare *single, thribble,* and *fourble.*

double board *n:* the name used for the working platform of the derrickman, or monkeyboard, when it is located at a height in the derrick or mast equal to two lengths of pipe joined together. Compare *fourble board* and *thribble board.*

double-drum hoist *n:* a device, consisting of two reels on which wire rope is wound, that provides two hoisting drums in the assembly. The main drum is used for pulling tubing or drill pipe; the second drum is used for coring and swabbing. See *hoist.*

double extra-strong pipe *n:* a pipe incorporating twice as many safety design factors as normally used.

double-pole mast *n:* a well-servicing unit whose mast consists of two steel tubes. Double-pole masts provide racking platforms for handling rods and tubing in stands and extend from 65 to 67 feet (20 m) so that rods can be suspended as 50-foot (15 m) doubles and tubing set back as 30-foot (9 m) singles. See *pole mast.*

doughnut *n:* a ring of wedges or a threaded, tapered ring that supports a string of pipe.

dowel *n:* a pin fitting into a hole in an abutting piece to prevent motion or slipping.

downcomer *n:* a tube that conducts liquids downward in a vessel such as an absorber.

downhole *adj, adv:* pertaining to the wellbore.

downhole mud motor *n:* also called a turbodrill or Dyna-Drill. See *turbodrill* and *Dyna-Drill.*

downtime *n:* time during which rig operations are temporarily suspended because of repairs or maintenance.

dozer *n:* a powered machine for earthwork excavations; a bulldozer.

DP *abbr:* drill pipe; used in drilling reports.

draft *n:* the vertical distance between the bottom of a vessel floating in water and the waterline.

draft marks *n pl:* numbers placed on the sides or ends of a floating offshore drilling rig to indicate its draft. See *draft*.

drag bit *n:* also called fishtail bit. See *fishtail bit*.

drainage *n:* the migration of oil or gas in a reservoir toward a wellbore due to pressure reduction caused by the well's penetration of the reservoir. A drainage point is a wellbore (or in some cases several wellbores) that drains the reservoir.

drainage radius *n:* the area of a reservoir in which a single well serves as a point for drainage of reservoir fluids.

Drake well *n:* the first U.S. well drilled in search of oil. Some 69 feet deep, it was drilled near Titusville, Pa., and completed in 1859.

drawdown *n:* 1. the difference between static and flowing bottomhole pressures. 2. the distance between the static level and the pumping level of the fluid in the annulus of a pumping well.

drawworks *n:* the hoisting mechanism on a drilling rig. It is essentially a large winch that spools off or takes in the drilling line and thus raises or lowers the drill stem and bit.

drawworks-drum socket *n:* a receptacle on the drawworks drum by which the drilling line is attached.

dress *v:* to sharpen or repair items of equipment (such as drilling bits and tool joints).

drift *v:* to move slowly out of alignment, off center, or out of register. 2. to gauge or measure pipe by means of a mandrel passed through it to ensure the passage of tools, pumps, and so forth.

drift angle *n:* also called angle of deviation and angle of drift. See *deviation*.

drift diameter *n:* 1. in drilling, the effective hole size. 2. in casing, the guaranteed minimum diameter of the casing. The drift diameter is important because it indicates whether the casing is large enough for a specified size of bit to pass through.

drift indicator *n:* a device dropped or run down the drill stem on a wireline to a point just above the bit to measure the inclination of the well off vertical at that point. It does not measure the direction of the inclination.

drift survey *n:* See *deviation survey*.

drill *v:* to bore a hole in the earth, usually to find and remove subsurface formation fluids such as oil and gas.

drillable *adj:* pertaining to packers and other tools left in the wellbore to be broken up later by the drill bit. Drillable equipment is made of cast iron, aluminum, plastic, or other soft, brittle material.

drill ahead *v:* to continue drilling operations.

drill around *v:* 1. to deflect the wellbore away from an obstruction in the hole. 2. (slang) to get the better of someone (as, "He drilled around me and got the promotion").

drill bit *n:* the cutting or boring element used for drilling. See *bit*.

drill collar *n:* a heavy, thick-walled tube, usually steel, used between the drill pipe and the bit in the drill stem to provide a pendulum effect to the drill stem and weight to the bit.

drill collar sub *n:* a sub used between the drill string and the drill collars. See *sub*.

drill column *n:* also called the drill stem. See *drill stem*.

drilled show *n:* oil or gas in the mud circulated to the surface.

driller *n:* the employee directly in charge of a drilling or workover rig and crew. His main duty is operation of the drilling and hoisting equipment, but he is also responsible for downhole condition of the well, operation of downhole tools, and pipe measurements.

driller's BOP control panel *n:* a series of controls on the rig floor that the driller manipulates to open and close the blowout preventers.

driller's console *n:* a metal cabinet on the rig floor containing the controls that the driller manipulates to operate various components of the drilling rig.

driller's control panel *n:* also called driller's console. See *driller's console*.

driller's log *n:* a record that describes each formation encountered and lists the drilling time relative to depth, usually in 5- to 10-foot (1.5 to 3 m) intervals.

driller's method *n:* a well-killing method involving two complete and separate circulations; the first circulates the kick out of the well and the second circulates heavier mud through the wellbore.

driller's panel *n:* See *driller's console*.

driller's position *n:* the area immediately surrounding the driller's console. See *driller's console*.

driller's report *n:* a record kept on the rig for each tour to show the footage drilled, tests made on drilling fluid, bit record, and other noteworthy items.

drill floor *n:* also called rig floor or derrick floor. See *rig floor*.

drill in *v:* to penetrate the productive formation after the casing is set and cemented on top of the pay zone.

drilling barge *n:* See *barge*.

drilling block *n:* a lease or a number of leases of adjoining tracts of land that constitute a unit of acreage sufficient to justify the expense of drilling a wildcat.

drilling break *n:* a sudden increase in the rate of penetration by the drill bit. It sometimes indicates that the bit has penetrated a high-pressure zone and thus warns of the possibility of a blowout.

drilling contract *n:* an agreement made between a drilling company and an operating company to drill and complete a well, setting forth the obligation of each party, compensation, indentification, method of drilling, depth to be drilled, and so on.

drilling contractor *n:* an individual or group of individuals that own a drilling rig and contract their services for drilling wells.

drilling control *n:* a device that controls the rate of penetration on a bit by maintaining constant weight of a predetermined magnitude on the bit. It is also called an automatic driller or automatic drilling control unit.

drilling crew *n:* a driller, a derrickman, and two or more helpers who operate a drilling or workover rig for one tour each day.

drilling engine *n:* an internal-combustion engine used to power a drilling rig. From two to six engines are used on a rotary rig, fueled by diesel fuel, gasoline, or gas.

drilling engineer *n:* an engineer who specializes in the technical aspects of drilling.

drilling fluid *n:* circulating fluid, one function of which is to force cuttings out of the wellbore and to the surface. Other functions are to cool the bit and to counteract downhole formation pressure. While a mixture of barite, clay, water, and chemical additives is the most common drilling fluid, wells can also be drilled by using air, gas, water, or oil-base mud as the drilling fluid. See *mud*.

drilling fluid analysis *n:* See *mud analysis*.

drilling foreman *n:* the supervisor of drilling or workover operations on a rig. Also called a toolpusher, rig manager, rig supervisor, or rig superintendent.

drilling head *n:* a special rotating head that has a gear and pinion drive arrangement to allow turning of the kelly and simultaneous sealing of the kelly against well pressure.

drilling line *n:* a wire rope used to support the drilling tools. Also called the rotary line.

drilling mud *n:* a specially compounded liquid circulated through the wellbore during rotary drilling operations. See *mud*.

drilling pattern *n:* See *well spacing*.

drilling platform *n:* See *platform*.

drilling rate *n:* the speed with which the bit drills the formation; usually called the rate of penetration.

drilling recorder *n:* an instrument that records hook load, penetration rate, rotary speed and torque, pump rate and pressure, mud flow, and so forth during drilling.

drilling rig *n:* See *rig*.

drilling spool *n:* a fitting placed in the blowout preventer stack to provide space between preventers for facilitating stripping operations and to permit attachment of choke and kill lines for localizing possible erosion to the spool instead of to the more expensive pieces of equipment.

drilling superintendent *n:* an employee, usually of a drilling contractor, who is in charge of all drilling operations that the contractor is engaged in.

drilling template *n:* See *temporary guide base*.

drilling unit *n:* the acreage allocated to a well when a regulatory agency grants a well permit.

drill out *v:* to remove with the bit the residual cement that normally remains in the lower section of casing and the wellbore after the casing has been cemented.

drill pipe *n:* heavy seamless tubing used to rotate the bit and circulate the drilling fluid. Joints of pipe approximately 30 feet (9 m) long are coupled together by means of tool joints.

drill pipe cutter *n:* a tool to cut drill pipe stuck in the hole. Tools that cut the pipe either internally or externally, permitting some of it to be withdrawn, are available. Jet cutoff or chemical cutoff is also used to free stuck pipe.

drill pipe float *n:* a valve installed in the drill stem that allows mud to be pumped down the drill stem but prevents flow back up the drill stem; a check valve.

drill pipe pressure *n:* the amount of pressure exerted inside the drill pipe as a result of circulating pressure, entry of formation pressure into the well, or both.

drill pipe pressure gauge *n:* an indicator, mounted in the mud circulating system, that measures and indicates the amount of pressure in the drill stem. See *drill stem*.

drill pipe protector *n:* an antifriction device of rubber or steel attached to each joint of drill pipe to minimize wear.

drill pipe safety valve *n:* a special valve used to close off the drill pipe to prevent backflow during a kick. It has threads to match the drill pipe in use.

drill pipe slips *n pl:* wedge-shaped pieces of metal with gripping elements, used to hold drill pipe in place and keep it from slipping down into the hole. See *slips*.

drill ship *n:* a ship constructed to permit a well to be drilled from it at an offshore location. While not as stable as other floating structures (such as a semisubmersible rig), drill ships, or shipshapes, are capable of drilling exploratory wells in relatively deep waters. They may have a ship hull, a catamaran hull, or a trimaran hull.

drill site *n:* the location of a drilling rig.

drill stem *n:* all members in the assembly used for rotary drilling from the swivel to the bit, including the kelly, drill pipe and tool joints, drill collars, stabilizers, and various specialty items. Compare *drill string*.

drill stem test *n:* the conventional method of formation testing. The basic drill stem test tool consists of a packer or packers, valves or ports that may be opened and closed from the surface, and two or more pressure-recording devices. The tool is lowered on the drill string to the zone to be tested. The packer or packers are set to isolate the zone from the drilling fluid column. The valves or ports are then opened, to allow for formation flow while the recorders chart flow pressures, and are then closed, to shut in the formation while the recorders chart static pressures. A sampling chamber traps clean formation fluids at the end of the test. Analysis of the pressure charts is an important part of formation testing.

drill string *n:* the column, or string, of drill pipe with attached tool joints that transmits fluid and rotational power from the kelly to the drill collars and bit. Often, especially in the oil patch, the term is loosely applied to both drill pipe and drill collars. Compare *drill stem*.

drill to granite *v:* to drill a hole until basement rock is encountered, usually in a wildcat well. If no hydrocarbon-bearing formations are found above the basement, the well is assumed to be dry.

drill under pressure *v:* to carry on drilling operations while maintaining a seal (usually with a rotating head) to prevent the well fluids from blowing out. Drilling under pressure is advantageous in that the rate of penetration is relatively fast; however, the technique requires extreme caution.

drip *n:* 1. the water and hydrocarbon liquids that have condensed from the vapor state in the natural gas flow line and accumulated in the low points of the line. 2. the receiving vessel that accumulates such liquids.

drip accumulator *n:* the device used to collect liquid hydrocarbons that condense out of a wet gas traveling through a pipeline.

drip gasoline *n:* hydrocarbon liquid that separates in a pipeline transporting gas from the well casing, lease separation, or other facilities and drains into equipment from which the liquid can be removed.

drive *n:* the means by which a machine is given motion or power, or by which power is transferred from one part of a machine to another. *v:* to give motion or power.

drive bushing *n:* also called kelly bushing. See *kelly bushing*.

drive chain *n:* a chain by means of which a machine is propelled.

drive-in unit *n:* a type of portable servicing or workover rig that is self-propelled, using power from the hoisting engines. The driver's cab and steering wheel are mounted on the same end as the mast support; thus the unit can be driven straight ahead to reach the wellhead. See *carrier rig*.

drive rollers *n:* also called kelly bushing rollers. See *kelly bushing rollers*.

drive shaft *n:* a shaft that transmits mechanical power to drive a machine.

drlg *abbr:* drilling; used in drilling reports.

drum *n:* 1. a cylinder around which wire rope is wound in the drawworks. The drawworks drum is that part of the hoist upon which the drilling line is wound. 2. a steel container of general cylindrical form. Refined products are shipped in steel drums with capacities of about 50 to 55 U.S. gallons or about 200 litres.

drum brake *n:* a device for arresting the motion of a machine or mechanism by means of mechanical friction; in this case, a shoe is pressed against a turning drum.

dry bed *n:* the solid adsorption materials such as molecular sieves, charcoal, or other materials used for purifying or for recovering liquid from a gas. See *adsorption*.

dry-bed dehydrator *n:* a connection of devices for removing water from gas in which two or more beds of solid desiccant are used. The wet gas is sent through one bed for drying while the other is prepared for later use.

dry drilling *n:* a drilling operation in which no fluid is circulated back up to the surface (often resulting in lost circulation). However, the fluid is usually circulated into the well to cool the bit. See *blind drilling*.

dry gas *n:* 1. gas whose water content has been reduced by a dehydration process. 2. gas containing few or no hydrocarbons commercially recoverable as liquid product. Also called lean gas.

dry hole *n:* any well that does not produce oil or gas in commercial quantities. A dry hole may flow water, gas, or even oil, but not enough to justify production.

dry oil *n:* oil that has been treated so that only small quantities of water and other extraneous materials remain in it.

dry string *n:* the drill pipe from which drilling mud has been emptied as it is pulled out of the wellbore.

dry suit *n:* a protective diving garment that is completely sealed to prevent water entry. It is designed to accommodate a layer of insulation between the diver and the suit. This insulation gives a diver maximum warmth and protection from the water.

dry welding *n:* arc, gas, or plasma welding performed in an underwater habitat with a gas environment at ambient pressure.

dry well *n:* See *dry hole*.

DST *abbr:* drill stem test.

dual completion *n:* a single well that produces from two separate formations at the same time. Production from each zone is segregated by running two tubing strings with packers inside the single string of production casing, or by running one tubing string with a packer through one zone while the other is produced through the annulus. In a miniaturized dual completion, two separate 4½-inch or smaller casing strings are run and cemented in the same wellbore.

dual induction focused log *n:* a log designed to provide the resistivity measurements necessary to estimate the effects of invasion so that more reliable values for the true formation resistivity may be obtained. The resistivity curves on this log are made by a deep, a medium, and a shallow investigation induction. Visual observation of the dual induction focused log can give valuable information regarding invasion, porosity, and hydrocarbon content. The three curves on the log can be used to correct for deep invasion and obtain a better value for formation resistivity. See *induction survey*.

duck's nest *n:* a relatively small, excavated earthen pit, into which are channeled excess quantities of drilling mud that overflow the usual pits.

ductile *adj:* capable of being permanently drawn out without breaking (e.g., wire may be ductile).

dump *n:* the volume of oil delivered to a pipeline in a complete cycle of a measuring tank in a LACT installation. A series of such dumps covered by a single scheduling ticket is called a run.

dump bailer *n:* a bailing device with a release valve, usually of the disk or flapper type, used to place or spot material (such as cement slurry) at the bottom of the well.

dump meter *n:* a liquid-measuring device consisting of a small tank with narrowed sections at top and bottom that automatically fills and empties itself and records the number of dumps.

dump tank *n:* a calibrated metering tank designed to automatically release an accurate volume of liquid; also called a measuring tank.

dump valve *n:* the discharge valve through which oil and water are discharged from separators, treaters, and so on. It is usually a motor valve, but may be a liquid-level controller as well.

duplex pump *n:* a reciprocating pump having two pistons or plungers, used extensively as a mud pump on drilling rigs.

duster *n:* a dry hole.

dutchman *n:* 1. the portion of a stud or screw that remains in place after the head has inadvertently been twisted off. 2. a tool joint pin broken off in the drill pipe box or drill collar box.

dwt *abbr:* deadweight ton.

Dyna-Drill *n:* a downhole motor driven by drilling fluid that imparts rotary motion to a drilling bit connected to the tool, thus eliminating the need to turn the entire drill stem to make hole. The Dyna-Drill, a trade name, is used in straight and directional drilling.

Dynaflex tool n: the trade name for a directional drilling tool that deflects the drilling assembly off vertical without having to be pulled from the hole. The device that causes the tool to be deflected can be caught and retrieved with a wireline.

dynamic loading n: exerting force with continuous movement; cyclic stressing.

dynamic positioning n: a method by which a floating offshore drilling rig is maintained in position over an offshore well location. Generally, several propulsion units, called thrusters, are located on the hulls of the structure and are actuated by a sensing system. A computer to which the system feeds signals directs the thrusters to maintain the rig on location.

dynamometer n: 1. a device used to measure the brake horsepower of a prime mover. 2. in sucker rod pumping, a device used to indicate a variation in load on the polished rod as the rod string reciprocates. A continuous record of the result of forces acting along the axis of the polished rod is provided on a dynamometer card, from which an analysis is made of the performance of the well pumping equipment.

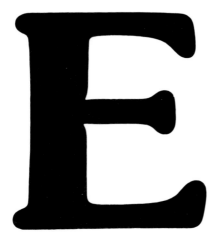

easement *n:* a right that one individual or company has on another's land. In the petroleum industry, it usually refers to the permission given by a landowner for a pipeline or access road to be laid across his property.

ebullition *n:* boiling, as especially applied to a system to remove heat from engine jacket water, wherein the water is permitted to boil and the evolved vapors are condensed in air-fin coolers.

ecology *n:* the study of living things and their relation to their environment.

edgewater *n:* the water that touches the edge of the oil in the lower horizon of a formation.

edge well *n:* a well on the outer fringes of a productive subsurface formation.

eductor *n:* a device similar to an ejector for mixing two fluids.

effective permeability *n:* a measure of the ability of a single fluid to flow through a rock when the pore spaces of the rock are not completely filled or saturated with the fluid. Compare *absolute permeability* and *relative permeability*.

effective porosity *n:* the percentage of the bulk volume of a rock sample that is composed of interconnected pore spaces that allow the passage of fluids through the sample. See *porosity*.

elastic collision *n:* a collision between a neutron and the nucleus of an atom of an element such as hydrogen. In such a collision, the neutron's energy is reduced by exactly the amount transferred to the nucleus with which it collided. The angle of collision and the relative mass of the nucleus determine energy loss.

elbow *n:* a fitting that allows two pipes to be joined together at an angle of less than 180 degrees, usually 90 degrees or 45 degrees.

elec log *abbr:* electric log; used in drilling reports.

electrical potential *n:* voltage.

electric dehydration *n:* See *emulsion breaker*.

electric drive *n:* See *electric rig*.

electric-drive rig *n:* also called an electric rig. See *electric rig*.

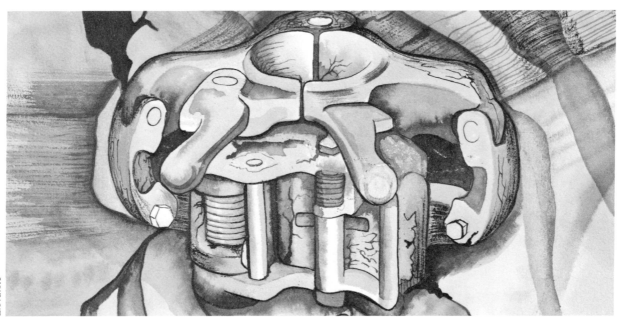

Elevators

electric generator *n:* a machine by which mechanical energy is changed into electrical energy, as an electric generator on a drilling rig in which a diesel engine (mechanical power) turns a generator to make electricity (electrical energy).

electrician *n:* the rig crew member who maintains and repairs the electrical generation and distribution system on the rig.

electric line *n:* common field term for conductor line. See *conductor line*.

electric log *n:* also called an electric well log. See *electric well log*.

electric rig *n:* a drilling rig on which the energy from the power source is distributed to the various rig components through electrical conductors, as opposed to being distributed by mechanical transmission. Such a rig has an electric drive.

electric survey *n:* also called an electric well log. See *electric well log*.

electric well log *n:* a record of certain electrical characteristics of formations traversed by the borehole, made to identify the formations, determine the nature and amount of fluids they contain, and estimate their depth. Also called an electric log or electric survey.

electrochemical series *n:* See *electromotive series*.

electrochemical treater *n:* See *electrostatic treater*.

electrode *n:* a conductor of electric current as it leaves or enters a medium such as an electrolyte, a gas, or a vacuum.

electrodeposition *n:* an electrochemical process by which metal settles out of an electrolyte that contains the metal's ions and is then deposited at the cathode of the cell.

electrodynamic brake *n:* a device mounted on the end of the drawworks shaft of a drilling rig. The electrodynamic brake (sometimes called a magnetic brake) serves as an auxiliary to the mechanical brake when pipe is lowered into a well. The braking effect in an electrodynamic brake is achieved by means of the interaction of electric currents with magnets, with other currents, or with themselves.

electrolysis *n:* the decomposition of a chemical compound brought about by the passage of an electrical current through the compound or through the solution containing the compound. Corroding action of stray currents is caused by electrolysis.

electrolyte *n:* 1. a chemical that, when dissolved in water, dissociates into positive and negative ions, thus increasing its electrical conductivity. See *dissociation*. 2. the electrically conductive solution that must be present for a corrosion cell to exist.

electrolytic property *n:* the ability of a substance, usually in solution, to conduct an electric current.

electromotive force *n:* 1. the force that drives electrons and thus produces an electric current. 2. the voltage or electric pressure that causes an electric current to flow along a conductor. The abbreviation for electromotive force is emf.

electromotive series *n:* a list of elements arranged in order of activity (tendency to lose electrons). The following metals are so arranged: magnesium, beryllium, aluminum, zinc, chromium, iron, cadmium, nickel, tin, copper, silver, and gold. If two metals widely separated in the list (e.g., magnesium and iron) are placed in an electrolyte and connected by a metallic conductor, an electromotive force is produced. See *corrosion*.

electron *n:* a particle in an atom that has a negative charge. An atom contains the same number of electrons and protons (which have a positive charge). Electrons orbit the nucleus of the atom.

electrostatic treater *n:* a vessel that receives emulsion and resolves the emulsion to oil, water, and usually gas, by using heat, chemicals, and a high-voltage electric field. This field, produced by grids placed perpendicular to the flow of fluids in the treater, aids in breaking the emulsion. Also called an electrochemical treater. See *emulsion treating*.

element *n:* one of more than a hundred simple substances that consist of atoms of only one kind and that either singly or in combination makes up all matter. For example, the simplest element is hydrogen, and one of the most abundant elements is carbon. Some elements such as radium and uranium are radioactive.

elev *abbr:* elevation; used in drilling reports.

elevated tank *n:* a vessel above a datum line (usually ground level).

elevator bails *n pl:* also called elevator links. See *elevator links*.

elevator links *n pl:* cylindrical bars that support the elevators and attach them to the hook. Also called elevator bails.

elevators *n pl:* a set of clamps that grips a stand of casing, tubing, drill pipe, or sucker rods so that the stand can be raised from or lowered into the hole.

elliptical tank *n:* a tank that has an elliptical cross section.

emf *abbr:* electromotive force.

emulsified water *n:* water so thoroughly combined with oil that special treating methods must be applied to separate it from the oil. Compare *free water.*

emulsifying agent *n:* material added to solid-in-liquid or liquid-in-liquid suspensions to separate the individual suspended particles. Also called a disperser or an emulsifier.

emulsion *n:* a mixture in which one liquid, termed the dispersed phase, is uniformly distributed (usually as minute globules) in another liquid, called the continuous phase or dispersion medium. In an oil-water emulsion, the oil is the dispersed phase and the water the dispersion medium; in a water-oil emulsion, the reverse holds. A typical product of oilwells, water-oil emulsion is also used as a drilling fluid.

emulsion breaker *n:* a system, device, or process used for breaking down an emulsion and producing two or more easily separated compounds (such as water and oil). Emulsion breakers may be (1) devices to heat the emulsion, thus achieving separation by lowering the viscosity of the emulsion and allowing the water to settle out; (2) chemical compounds, which destroy or weaken the film around each globule of water, thus uniting all the drops; (3) mechanical devices such as settling tanks and wash tanks; or (4) electrostatic treaters, which use an electric field to cause coalescence of the water globules.

emulsion test *n:* a procedure carried out to determine the proportions of sediment and dispersed compounds in an emulsion. Such tests may range from elaborate distillation conducted in laboratories to simple and expedient practices used in the field.

emulsion treating *n:* the process of breaking down emulsions to separate oil from water or other contaminants. Treating plants may use a single process or a combination of processes to effect demulsification, depending upon what emulsion is being treated.

encroachment *n:* See *water encroachment.*

Energy Telecommunications and Electrical Association *n:* a Houston-based, nonprofit, tax-exempt organization that aids the energy industry in dealing with technological advances by advising and educating its members on new equipment and safety.

engine *n:* a machine for converting the heat content of fuel into rotary motion that can be used to power other machines. An engine is not a motor.

Engler distillation *n:* a test that determines the volatility of a gasoline by measuring the percentage of the gasoline that can be distilled at various temperatures.

enhanced oil recovery *n:* 1. the introduction of an artificial drive and displacement mechanism into a reservoir to produce oil unrecoverable by primary recovery methods. The purpose of enhanced oil recovery (EOR) is to restore formation pressure and fluid flow to a substantial portion of a reservoir by injecting fluids into injection wells located in rock that has fluid communication with production wells. EOR methods include waterflooding, chemical flooding, gas injection, and thermal recovery. Chemical flooding, most types of gas injection, and thermal methods are often called advanced EOR methods because they not only restore formation pressure but also improve displacement of oil by overcoming forces that keep the oil trapped in rock pores. 2. the use of an advanced EOR method. See *primary recovery, secondary recovery, tertiary recovery,* and the various EOR methods.

ENTELEC *abbr:* Energy Telecommunications and Electrical Association.

enthalpy *n:* the heat content of fuel. A thermodynamic property, it is the sum of the internal energy of a body and the product of its pressure multiplied by its volume.

entrained gas *n:* formation gas that enters the drilling fluid in the annulus. See *gas-cut mud.*

entrained liquid *n:* liquid particles that may be carried out of the top of a distillation or absorber column with the vapors or residue gas.

entropy *n:* the internal energy of a substance that is attributed to the internal motion of the molecules. This energy is within the molecules and cannot be utilized for external work.

Environmental Protection Agency *n:* a federal agency for the supervision and control of the environmental quality.

environment of deposition *n:* See *depositional environment.*

EPA *abbr:* Environmental Protection Agency.

epithermal neutron *n:* a neutron having an energy level of 0.02 to 100 ev.

epithermal neutron log *n:* a log designed to have maximum sensitivity to detectable neutrons with energies above the thermal level and minimal sensitivity to capture gamma rays and thermal neutrons. Since an epithermal neutron log is only slightly affected by capture cross section and capture gamma ray emission, it reduces errors that arise from variations in formation chemistry.

EP mix *abbr:* ethane-propane mix.

epoxy *n:* any compound characterized by the presence of a reactive chemical structure that has an oxygen atom joined to each of two carbon atoms that are already bonded.

equalizer *n:* a device used with mechanical brakes of a drawworks to ensure that, when the brakes are used, each brake band will receive an equal amount of tension and also that, in case one brake fails, the other will carry the load. A mechanical brake equalizer is a dead anchor attached to the drawworks frame in the form of a yoke attached to each brake band. Some drawworks are equipped with automatic equalizers.

equilibrium constant *n:* See *vapor-liquid equilibrium ratio.*

erosion *n:* the process by which material (such as rock or soil) is worn away or removed (as by wind or water).

erosion drilling *n:* the high-velocity ejection of a stream of drilling fluid from the nozzles of a jet bit to remove rock encountered during drilling. Sometimes sand or steel shot is added to the drilling fluid to increase its erosive capabilities.

ES *abbr:* electric survey. See *electric well log.*

escarpment *n:* a cliff or relatively steep slope that separates level or gently sloping areas of land.

est *abbr:* estimated; used in drilling reports.

estimated ROB (remaining on board) *n:* estimated material remaining on board a vessel after a discharge. Includes residue or sediment clingage, which builds up on the interior surfaces of the vessel's cargo compartments.

ethane *n:* a light hydrocarbon, C_2H_6, found in natural gas. It is a gas under atmospheric conditions.

ethylene *n:* a chemical compound of the olefin series having the formula C_2H_4. Official name is ethene.

ethylene glycol *n:* a colorless liquid used as an antifreeze and as a dehydration medium in removing water from gas. See *glycol dehydrator.*

evaporation loss *n:* a loss to the atmosphere of petroleum fractions through evaporation, usually while the fractions are in storage or in process. See *vaporization.*

evaporator *n:* a vessel used to convert a liquid into its vapor phase.

evaporite *n:* a sedimentary rock (such as gypsum or salt) that originates from the evaporation of seawater in enclosed basins.

evening tour *n:* (pronounced "tower") the shift of duty on a drilling rig that starts in the afternoon and runs through the evening. Compare *daylight tour* and *graveyard tour.*

even keel *n:* on a ship or floating offshore drilling rig, the balance when the plane of flotation is parallel to the keel.

excelsior *n:* a fibrous material used as a filtering element in the hay section of heaters or heater-treaters.

exchanger *n:* a piping arrangement that permits heat from one fluid to be transferred to another fluid as they travel countercurrent to one another. In the heat exchanger of an emulsion-treating unit, heat from the outgoing clean oil is transferred to the incoming well fluid, cooling the oil and heating the well fluid.

excited state of nucleus *n:* the increased energy condition of the nucleus of an atom after it has captured a neutron. The nucleus becomes intensely excited and begins emitting a high-energy gamma ray called a capture gamma ray.

exciter *n:* a small DC generator mounted on top of a main generator to produce the field for the main generator.

exhaust *v:* to remove the burned gases from the cylinder of an engine. *n:* the burned gases that are removed from the cylinder of an engine.

exhaust manifold *n:* a piping arrangement, immediately adjacent to the engine, that collects burned gases from the engine and channels them to the exhaust pipe.

exhaust valve *n:* the cam-operated mechanism through which burned gases are ejected from an engine cylinder.

expanded perlite *n:* a siliceous volcanic rock that is finely ground and subjected to extreme heat. The resulting release of water leaves the rock particles considerably expanded and thus more porous. Expanded perlite is sometimes used in cement to increase its yield and decrease its density without an appreciable effect on its other properties.

expanding cement *n:* cement that expands as it sets to form a tighter fit around casing and formation.

expansion dome *n:* a cylindrical projection on top of a tank, tank car, or truck, into which liquids may expand without overflowing. Gauge point is often in the expansion dome. Shell thickness is part of dome capacity.

expansion joint n: a device used to connect long lines of pipe to allow the pipe joints to expand or contract as the temperature rises or falls.

expansion loop n: a full loop built into a pipeline to allow for expansion and contraction of the line.

expansion refrigeration n: cooling obtained from the evaporation of a liquid refrigerant or the expansion of a gas.

expansion turbine n: a device that converts the energy of a gas or vapor stream into mechanical work by expanding the gas or vapor through a turbine.

expendable gun n: a perforating gun that consists of a metal strip upon which are mounted shaped charges in special capsules. After firing, nothing remains of the gun but debris. See gun-perforate.

expendable-retrievable gun n: a perforating gun that consists of a hollow, cylindrical carrier, into which are placed shaped charges. Upon detonation, debris created by the exploded charges falls into the carrier and is retrieved when the gun is pulled out of the hole; however, the gun cannot be reused. See gun-perforate.

exploitation n: the development of a reservoir to extract its oil.

exploitation well n: a well drilled to permit more effective extraction of oil from a reservoir. Sometimes called a development well. See development well.

exploration n: the search for reservoirs of oil and gas, including aerial and geophysical surveys, geological studies, core testing, and drilling of wildcats.

exploration well n: also called a wildcat. See wildcat.

explosimeter n: an instrument used to measure the concentration of combustible gases in the air. Also called a gas sniffer.

extender n: 1. a substance added to drilling mud to increase viscosity without adding clay or other thickening material. 2. an additive that assists in getting a greater yield from a sack of cement. The extender acts by requiring more water than that required by neat cement.

external cutter n: a fishing tool, containing metal-cutting knives, that is lowered into the hole and over the outside of a length of pipe to cut it. The severed portion of the pipe can then be brought to the surface. Compare internal cutter.

external phase n: See continuous phase.

external upset n: an extra-thick wall on the outside of the threaded end of drill pipe or tubing. Externally upset pipe does not have a uniform diameter throughout its length but is enlarged at each end. Compare internal upset.

extraction n: the process of separating one material from another by means of a solvent. The term can be applied to absorption, liquid-liquid extraction, or any other process using a solvent.

extraction plant n: a plant equipped to remove liquid constituents from casinghead gas or wet gas.

extrusive rocks n: igneous rocks formed from lava poured out on the earth's surface.

F *abbr:* Fahrenheit. See *Fahrenheit scale*.

F *sym:* farad.

face mask *n:* a mask used to seal all or a portion of a diver's face from the underwater environment, made of a rubber frame surrounding a clear flat lens.

facies *n:* part of a bed of sedimentary rock that differs significantly from other parts of the bed.

Fahrenheit scale *n:* a temperature scale devised by Gabriel Fahrenheit, in which 32 degrees represents the freezing point and 212 degrees the boiling point of water at standard sea-level pressure. Fahrenheit degrees may be converted to Celsius degrees by using the following formula:

$$C = {}^5/_9 (F - 32)$$

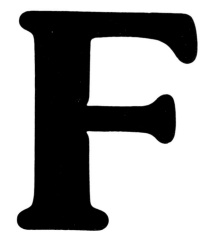

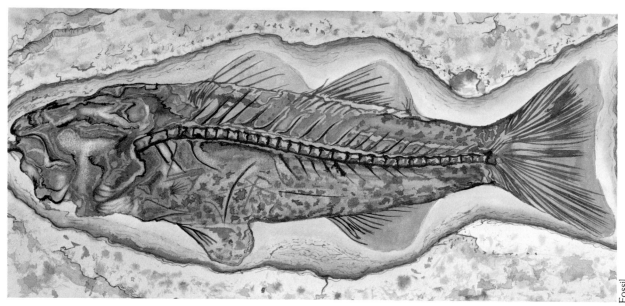

Fossil

fairleader *n:* a device used to maintain or alter the direction of a rope or a chain so that it leads directly to a sheave or drum without undue friction.

Fann V-G meter *n:* trade name of a device used to record and measure at different speeds the flow properties of plastic fluids (such as the viscosity and gel strength of drilling fluids).

farad *n:* the metric unit of electrical capacitance. Its symbol is F.

farm boss *n:* an oil company supervisor who controls production activities within a limited area.

farm-in *n:* a contract identical to a farmout, with the operator as the third party. See *farmout*.

farm in *v:* to accept, as an operator, the third-party assignment of a farmout. See *farmout*.

farmout *n:* a contract between a lessee and a third party to assign leasehold interest to the third party, conditional upon the third party's drilling a well within the expiration date of the primary term of the lease. The assignment may include the entire interest together with dry-hole money, or partial interest or entire interest with an override, typically $1/8$ to $1/12$. In the latter, after payout the override may convert to a 25% working interest. A farmout is distinguished from a joint operating agreement by the fact that the lessee does not incur any of the drilling costs.

farm out *v:* to assign leasehold interest to a third party, with stipulated conditions. See *farmout*.

fastline *n:* the end of the drilling line that is affixed to the drum or reel of the drawworks, so called because it travels with greater velocity than any other portion of the line. Compare *deadline*.

fast sheave *n:* that sheave on the crown block over which the fastline is reeved.

fathom *n:* a measure of ocean depth in countries using the English system of measurement, equal to 6 feet or 1.83 m.

fatigue *n:* the tendency of a material such as a metal to break under repeated cyclic loading at a stress considerably less than the tensile strength shown in a static test.

fault *n:* a break in subsurface strata. Often strata on one side of the fault line have been displaced (upward, downward, or laterally) relative to their original positions.

fault plane *n:* a surface along which faulting has occurred.

fault trap *n:* a subsurface hydrocarbon trap created by faulting, which causes an impermeable rock layer to be moved opposite the reservoir bed.

feed off *v:* to lower the bit continuously or intermittently by allowing the brake to disengage and the drum to turn. The feed-off rate is the speed with which the cable is unwound from the drum.

feed tank *n:* a vessel containing a charge stock or a vessel from which a stream is continuously fed for further processing.

female connection *n:* a pipe, coupling, or tool threaded on the inside so that only a male connection can be joined to it.

FeO *form:* ferrous oxide.

ferrous oxide *n:* a black, powdery compound of iron and oxygen (FeO) produced in petroleum operations from the oxidation of ferrous sulfide (FeS_2), a reaction that releases a great deal of heat. Also called iron monoxide.

ferrous sulfide *n:* a black crystalline compound of iron and sulfur (FeS_2) produced in petroleum operations from the reaction of hydrogen sulfide (H_2S) and the iron (Fe) in steel. Also called iron sulfide.

FeS$_2$ *form:* ferrous sulfide.

fibrous material *n:* any tough, stringy material of threadlike structure used to prevent loss of circulation or to restore circulation in porous or fractured formations.

field *n:* a geographical area in which a number of oil or gas wells produce from a continuous reservoir. A field may refer to surface area only or to underground productive formations as well. A single field may have several separate reservoirs at varying depths.

field-grade butane *n:* a product consisting chiefly of normal butane and isobutane, produced at a gas processing plant. Also called mixed butane.

field processing *n:* the processing of oil and gas in the field before delivery to a major refinery or gas plant, including separation of oil from gas, separation of water from oil and from gas, and removal of liquid hydrocarbons.

field processing unit *n:* a unit through which a well stream passes before the gas reaches a processing plant or sales point.

field pump *n:* a pump installed in a field to transfer oil from a lease tank to a central gathering station near a main pipeline.

field superintendent *n:* an employee of an oil company who is in charge of a particular oil or gas field from which the company is producing.

FIH *abbr:* fluid in hole; used in drilling reports.

filler material *n:* a material added to cement or cement slurry to increase its yield.

fill the hole *v:* to pump drilling fluid into the wellbore while the pipe is being withdrawn in order to ensure that the wellbore remains full or fluid even though the pipe is withdrawn. Filling the hole lessens the danger of blowout or of caving of the wall of the wellbore.

fill-up line *n:* the smaller of the side fittings on a bell nipple, used to fill the hole when drill pipe is being removed from the well.

filter *n:* a porous medium through which a fluid is passed to separate particles of suspended solids from it.

filter cake *n:* 1. compacted solid or semisolid material remaining on a filter after pressure filtration of mud with a standard filter press. Thickness of the cake is reported in thirty-seconds of an inch or in millimetres. 2. the layer of concentrated solids from the drilling mud or cement slurry that forms on the walls of the borehole opposite permeable formations; also called wall cake or mud cake.

filter loss *n:* the amount of fluid that can be delivered through a permeable filter medium after being subjected to a set differential pressure for a set length of time.

filter press *n:* a device used in the testing of filtration properties of drilling mud. See *mud*.

filtrate *n:* a fluid that has been passed through a filter.

filtration *n:* the process of filtering a fluid.

filtration loss *n:* the escape of the liquid part of a drilling mud into permeable formations.

fin *n:* a thin sharp ridge around the box or the pin shoulder of a tool joint, caused by the use of boxes and pins with different-sized shoulders.

final squeeze pressure *n:* the fluid pressure at the completion of a squeeze-cementing operation.

fine *n:* a fragment or particle of rock or mineral that is too minute to be treated as ordinary coarse material.

fingerboard *n:* a rack that supports the tops of the stands of pipe being stacked in the derrick or mast. It has several steel fingerlike projections that form a series of slots into which the derrickman can set a stand of drill pipe after it is pulled out of the hole and removed from the drill string.

fingering *n:* a phenomenon that often occurs in an injection well in which the fluid being injected does not contact the entire reservoir but bypasses sections of the reservoir fluids in a fingerlike manner. Fingering is not desirable because portions of the reservoir are not contacted by the injection fluid.

fins *n:* semirigid, paddlelike extensions worn on the feet to increase propulsion power while swimming.

fin-tube *n:* a tube or pipe having an extended surface in the form of fins used in heat exchangers or other heat-transfer equipment.

fire *v:* to start and maintain the fire in a boiler or heater.

fired heater *n:* a furnace in which natural gas or other fuel is burned to heat the gas or liquid passing through the furnace tubes.

fire flood *n:* also called *in situ* combustion. See *in situ combustion.*

fireman *n:* the member of the crew on a steam-powered rig who is responsible for care and operation of the boilers. Compare *motorman.*

fire point *n:* the temperature at which a petroleum product burns continuously after being ignited. See *flash point.*

fire tube *n:* A pipe, or set of pipes, within an emulsion settling tank through which steam or hot gases are passed to warm the emulsion and make the oil less viscous. See *steam coil.*

fire wall *n:* a structure erected to contain petroleum or a petroleum-fed fire in case a storage vessel ruptures or collapses. Usually a dike is built around a petroleum storage tank and a steel or stone wall put up between the prime movers and the oil pumps in a pipeline pumping station.

fish *n:* an object that is left in the wellbore during drilling or workover operations and that must be recovered before work can proceed. It can be anything from a piece of scrap metal to a part of the drill stem. *v:* 1. to recover from a well any equipment left there during drilling operations, such as a lost bit or drill collar or part of the drill string. 2. to remove from an older well certain pieces of equipment (such as packers, liners, or screen pipe) to allow reconditioning of the well.

fishing head *n:* a specialized fixture on a downhole tool that will allow the tool to be fished out after its use downhole. See *fish.*

fishing magnet *n:* a powerful permanent magnet designed to recover metallic objects lost in a well.

fishing tap *n:* a tap that goes inside pipe lost in a well to provide a firm grip and permit recovery of the fish; sometimes used in place of a spear.

fishing tool *n:* a tool designed to recover equipment lost in a well.

fishing-tool operator *n:* the person (usually a service company employee) in charge of directing fishing operations. See *fishing tool.*

fishtail bit *n:* a drilling bit with cutting edges of hard alloys; also called a drag bit. First used when the rotary system of drilling was developed about 1900, it is still useful in drilling very soft formations.

fissure *n:* a crack or fracture in a subsurface formation.

fitting *n:* a small, often standardized part (such as a coupling, valve, or gauge) installed into a larger apparatus.

fixed choke *n:* a choke whose opening is one size only; its opening is not adjustable. Compare *adjustable choke.*

fl/ *abbr:* flowed or flowing; used in drilling reports.

flag *n:* 1. a piece of cloth, rope, or nylon strand used to mark the wireline when swabbing or bailing. 2. an indicator of wind direction used during drilling or workover operations where hydrogen sulfide (sour) gas may be encountered. *v:* 1. to signal or attract attention. 2. in swabbing or bailing, to attach a piece of cloth to the wireline to enable the operator to estimate the position of the swab or bailer in the well.

flammable *adj:* capable of being easily ignited. Sometimes the term *inflammable* is used, but *flammable* is preferred because it correctly describes the condition.

flange *n:* a projecting rim or edge (as on pipe fittings and openings in pumps and vessels), usually drilled with holes to allow bolting to other flanged fittings.

flanged orifice fitting *n:* a two-piece orifice fitting with flanged faces that are bolted together.

flange union *n:* a device in which two matching flanges are used to join the ends of two sections of pipe.

flange up *v:* 1. to use flanges to make final connections on a piping system. 2. (slang) to complete any operation, as, "They flanged up the meeting and went home."

flare *n:* an arrangement of piping and burners used to dispose (by burning) of surplus combustible vapors, usually situated near a gasoline plant, refinery, or producing well. *v:* to dispose of surplus combustible vapors by igniting them in the atmosphere. Currently, flaring is rarely used because of the high value of gas as well as the stringent air pollution controls.

flare gas *n:* gas or vapor that is flared.

flash point *n:* the temperature at which a petroleum product ignites momentarily but does not burn continuously. Compare *fire point*.

flash set *n:* a premature thickening or setting of cement slurry, which makes it unpumpable.

flash tank *n:* a vessel used for separating the liquid phase from the gaseous phase formed from a rise in temperature and/or a reduction of pressure on the flowing stream.

flash welding *n:* a form of resistance butt welding used to weld wide, thin members or members with irregular faces together, and tubing to tubing.

flat *n:* See *kelly flat*.

fleet angle *n:* the angle created by drilling line between the drawworks drum and the fast sheave. The line is parallel to the sheave groove at only one point on the drum, and as the rope moves from this point either way, the fleet angle is created. The fleet angle should be held to a minimum – less than 1.5 degrees for grooved drums.

flex joint *n:* a device that provides a flexible connection between the riser pipe and the subsea blowout preventers. By accommodating lateral movement of a mobile offshore drilling rig, flex joints help to prevent a buildup of abnormal bending load pressure.

float *n:* an element of a level-control assembly designed to operate while partially or completely submerged in a liquid, the level of which is controlled by the assembly. The buoyancy of the liquid activates the float and the control valve to which it is linked and modifies the rate of the inflow or the outflow of the vessel to maintain a preset level. Sometimes a drill pipe float is called simply a float.

float collar *n:* a special coupling device, inserted one or two joints above the bottom of the casing string, that contains a check valve to permit fluid to pass downward but not upward through the casing. The float collar prevents drilling mud from entering the casing while it is being lowered, allowing the casing to float during its descent and thus decreasing the load on the derrick. A float collar also prevents backflow of cement during a cementing operation.

floater *n:* an offshore drilling structure that floats and is not secured to the seafloor (except for anchors). Semisubmersible drilling rigs and drill ships are floaters; platforms, submersible drilling rigs, and jackup drilling rigs are not. However, the latter two are floated before being positioned over the drilling site.

floating offshore drilling rig *n:* See *floater*.

floating roof *n:* a tank covering that rests on the surface of a hydrocarbon liquid in the tank and rises and falls as the liquid level rises and falls. Use of a floating roof eliminates vapor space above the liquid in the tank and conserves light fractions of the liquid.

floating tank *n:* a tank with its main gate valve open to the main line at a station. Oil may thus enter or leave the tank as pumping rates in the main line vary.

float shoe *n:* a short, heavy, cylindrical steel section with a rounded bottom, attached to the bottom of the casing string. It contains a check valve and functions similarly to the float collar but also serves as a guide shoe for the casing.

float switch *n:* a switch in a circuit that is opened or closed by the action of a float and that maintains a predetermined level of liquid in a vessel.

float valve *n:* a drill pipe float.

flocculation *n:* the coagulation of solids in a drilling fluid, produced by special additives or by contaminants.

flood *v:* 1. to drive oil from a reservoir into a well by injecting water under pressure into the reservoir formation. See *water flood*. 2. to drown out a well with water.

floodable length *n:* the length of a ship or mobile offshore drilling rig that may be flooded without its sinking below its safety or margin line, usually a few inches below the freeboard deck.

floor crew *n:* those workers on a drilling or workover rig who work primarily on the rig floor.

floorman *n:* also called a rotary helper. See *rotary helper*.

flotation cell *n:* a large, cylindrical tank in which water that is slightly oil-contaminated is circulated to be cleaned before it is disposed of overboard or into a disposal well. Since oil droplets cling to rapidly rising gas, usually a device such as a bubble tower is installed in the cell to permit the introduction of gas into the water.

flotation vest *n:* a vest most commonly worn by a sport diver to overcome the buoyancy effect of water and keep him afloat in the proper position. Carbon dioxide cartridges inside the vest are fired when inflation of the vest is necessary.

flow *n:* a current or stream of fluid.

flow bean *n:* a plug with a small hole drilled through it, placed in the flow line at a wellhead to restrict flow if it is too high. Compare *choke*.

flow by heads *v:* to produce intermittently.

flow chart *n:* a record, made by a recording meter, that shows the rate of production.

flowing pressure *n:* pressure registered at the wellhead of a flowing well.

flowing well *n:* a well that produces oil or gas by its own reservoir pressure rather than by use of artificial means (such as pumps).

flow line *n:* the surface pipe through which oil travels from a well to processing equipment or to storage.

flow-line treater *n:* a cylindrical vessel into which an emulsion is piped to be broken down into its components. See *heater-treater* and *electrostatic treater*.

flow-line treating *v:* the process of separating, or breaking down, an emulsion into oil and water in a vessel or tank on a continuous basis (i.e., without an interruption in the flow of emulsion into the tank or vessel). Compare *batch treating*.

flowmeter *n:* a device that measures the amount of fluid moving through a pipe. See *orifice meter* and *positive-displacement meter*.

flowstream *n:* the flow of fluids within a pipe.

flowstream samples *n:* small quantities of fluid taken at the wellhead or from the flow line and analyzed to determine composition of the flow.

flow tank *n:* also called a production tank. See *production tank*.

flow treater *n:* a single unit that acts as an oil and gas separator, an oil heater, and an oil and water-treating vessel; a heater-treater. See *heater-treater*.

fluid *n:* a substance that flows and yields to any force tending to change its shape. Liquids and gases are fluids.

fluid contact *n:* the approximate point in a reservoir where the gas-oil contact or oil-water contact is located.

fluid end *n:* the portion or end of a fluid pump that contains the parts involved in moving the fluid (such as liners and rods) as opposed to the end that produces the power for movement.

fluid level *n:* the distance from the earth's surface to the top of the liquid in the tubing or the casing in a well. The static fluid level is taken when the well is not producing and has stabilized. The dynamic, or pumping, level is the point to which the static level drops under producing conditions.

fluid loss *n:* the undesirable migration of the liquid part of the drilling mud or cement slurry into a formation, often minimized or prevented by the blending of additives with the mud.

fluid-loss additive *n:* a compound added to cement slurry or drilling mud to prevent or minimize fluid loss.

fluid pound *n:* the erratic impact of a pump plunger against a fluid when the pump is operating with a partial vacuum in the cylinder, with gas trapped in the cylinder, or with the well pumping off.

fluid sampler *n:* an automatic device that periodically takes a sample of a fluid flowing in a pipe.

fluid saturation *n:* the amount of the pore volume of a reservoir rock that is filled by water, oil, or gas and measured in routine core analysis.

flume *n:* also called a boot. See *boot*.

fluor *abbr:* fluorescence; used in drilling reports.

flush-joint casing *n:* a casing in which the outside diameter of the joint is the same as the outside diameter of the casing itself.

flush production *n:* a high rate of flow from a newly drilled well.

fluted drill collar *n:* See *spirally grooved drill collar.*

flywheel *n:* a large, circular disc, connected to and revolving with an engine crankshaft, that stores energy and disburses it as the engine runs.

fm *abbr:* formation; used in drilling reports.

foam drilling *n:* See *mist drilling.*

foaming agent *n:* a chemical used to lighten the water column in gas wells, in oilwells producing gas, and in drilling wells in which air or gas is used as the drilling fluid so that the water can be forced out with the air or gas to prevent its impeding the production or drilling rate. See *mist drilling.*

fold *n:* a flexure of rock strata into arches and troughs, produced by earth movements. See *anticline* and *syncline.*

footage rates *n:* a fee basis in drilling contracts stipulating that payment to the drilling contractor is made according to the number of feet of hole drilled.

foot-pound *n:* a unit of measure of work; the amount of energy required to move 1 pound 1 foot vertically. The metric equivalent is the centimetre-kilogram, or, in SI units, the joule.

foot valve *n:* a check valve at the inlet end of the suction pipe of a pump that enables the pump to remain full of liquid when it is not in operation.

forced draft *n:* air blown into a furnace or other equipment by a fan or blower.

fore and aft *n:* the lengthwise measurement of a mobile offshore drilling rig or ship.

formation *n:* a bed or deposit composed throughout of substantially the same kind of rock; a lithologic unit. Each different formation is given a name, frequently as a result of the study of the formation outcrop at the surface and sometimes based on fossils found in the formation.

formation boundary *n:* the horizontal limits of a formation.

formation damage *n:* the reduction of permeability in a reservoir rock caused by the invasion of drilling fluid and treating fluids to the section adjacent to the wellbore. It is often called skin. See *skin.*

Formation Density Log *n:* trade name for a density log.

formation dip *n:* the angle at which a formation bed inclines away from the horizontal.

formation evaluation *n:* the analysis of subsurface formation characteristics, such as lithology, porosity, permeability, and saturation, by indirect methods such as wireline well logging or by direct methods such as mud logging and core analysis.

formation fluid *n:* fluid (such as gas, oil, or water) that exists in a subsurface rock formation.

formation fracture pressure *n:* the point at which a formation will crack from pressure in the wellbore.

formation fracturing *n:* a method of stimulating production by increasing the permeability of the producing formation. Under extremely high hydraulic pressure, a fluid (such as water, oil, alcohol, dilute hydrocholoric acid, liquefied petroleum gas, or foam) is pumped downward through tubing or drill pipe and forced into the perforations in the casing. The fluid enters the formation and parts or fractures it. Sand grains, aluminum pellets, glass beads, or similar materials are carried in suspension by the fluid into the fractures. These are called propping agents or proppants. When the pressure is released at the surface, the fracturing fluid returns to the well, and the fractures partially close on the proppants, leaving channels for oil to flow through them to the well. This process if often called a frac job.

formation gas *n:* gas initially produced from an underground reservoir.

formation pressure *n:* the force exerted by fluids in a formation, recorded in the hole at the level of the formation with the well shut in. Also called reservoir pressure or shut-in bottomhole pressure.

formation strike *n:* the horizontal direction of a formation bed as measured at a right angle to the dip of the bed.

formation tester *n:* also called wireline formation tester. See *wireline formation tester.*

formation testing *n:* the gathering of pressure data and fluid samples from a formation to determine its production potential before choosing a completion method. Formation testing tools include formation testers and drill stem test tools.

formation water *n:* the water originally in place in a formation. See *connate water.*

formic acid *n:* a simple organic acid used for acidizing oilwells. It is stronger than acetic acid but much less corrosive than hydrofluoric or hydrochloric acid and is usually used for high-temperature wells.

forward *adv:* in the direction of the bow on a ship or an offshore drilling rig.

fossil *n:* the remains or impressions of a plant of past geological ages that have been preserved in or as rock.

fouling factor *n:* a factor used in heat-transfer calculations that represents the resistance to the flow of heat caused by dirt, scale, or other contaminants in the flowing fluids.

foundation pile *n:* the first casing or conductor string (generally with a diameter of 30 to 36 inches) set when drilling a well from an offshore drilling rig. It prevents sloughing of the ocean-floor formations and is a structural support for the permanent guide base and the blowout preventers.

fourble *n:* a section of drill pipe, casing, or tubing consisting of four joints screwed together. Compare *single*, *double*, and *thribble*.

fourble board *n:* the name used for the working platform of the derrickman, or the monkeyboard, when it is located at a height in the derrick equal to approximately four lengths of pipe joined together. Compare *double board* and *thribble board*.

four-stroke/cycle engine *n:* an engine in which the piston moves from top dead center to bottom dead center two times to complete a cycle of events. The crankshaft must make two complete revolutions, or 720 degrees.

four-way drag bit *n:* a drag bit with four blades. See *bit* and *fishtail bit*.

FP *abbr:* flowing pressure; used in drilling reports.

frac *abbr:* fractured; used in drilling reports.

frac job *n:* See *formation fracturing*.

fraction *n:* a part of a mixture of hydrocarbons, usually defined by boiling range—for example, naphtha, gas oil, kerosine, and others.

fractionate *v:* to separate single fractions from a mixture of hydrocarbon fluids, usually by distillation.

fractionating column *n:* the vessel or tower in a gas plant in which fractionation occurs. See *fractionate*.

fracture *n:* a crack or crevice in a formation, either natural or induced.

fracture acidizing *n:* a procedure by which acid is forced into a formation under pressure high enough to cause the formation to crack. The acid acts on certain kinds of rocks, usually carbonates, to increase the permeability of the formation. Compare *matrix acidizing*.

fracturing *n:* shortened form of formation fracturing. See *formation fracturing*.

free air space *n:* any of the cavities in the human body that contain air and are normally connected to the atmosphere, including lungs, sinuses, and middle ear.

freeboard *n:* the vertical distance between the waterline and the freeboard deck on a ship, boat, or floating offshore drilling rig. Draft plus freeboard equals total height of vessel.

free-point indicator *n:* a tool designed to measure the amount of stretch in a string of stuck pipe and to indicate the deepest point at which the pipe is free. The free-point indicator is lowered into the well on a conducting cable. Each end of a strain-gauge element is anchored to the pipe wall by friction springs or magnets, and, as increasing strain is put on the pipe, an accurate measurement of its stretch is transmitted to the surface. The stretch measurements indicate the depth at which the pipe is stuck.

free water *n:* 1. the water produced with oil. It usually settles out within five minutes when the well fluids become stationary in a settling space within a vessel. Compare *emulsified water*. 2. the measured volume of water that is present in a container and that is not in suspension in the contained liquid at observed temperature.

free-water knockout *n:* a vertical or horizontal vessel into which oil or emulsion is run in order to allow any water not emulsified with the oil (free water) to drop out.

freewheeling *n:* in an overrunning clutch, used between auxiliary hydrodynamic brake and the drawworks drumshaft, the action in which the clutch automatically disengages and runs freely while the empty block is being hoisted.

freeze point *n:* the depth in the hole at which the tubing, casing, or drill pipe is stuck. See *free-point indicator*.

freezing point *n:* the temperature at which a liquid becomes a solid.

fresh water *n:* water that has little or no salt dissolved in it.

freshwater mud *n:* See *mud*.

friction *n:* resistance to movement created when two surfaces are in contact. When friction is present, movement between the surfaces produces heat.

friction bearing *n: See journal bearing.*

friction clutch *n:* a clutch that makes connection by sliding friction.

friction loss *n:* a reduction in the pressure of a fluid caused by its motion against an enclosed surface (such as a pipe). As the fluid moves through the pipe, friction between the fluid and the pipe wall creates a pressure loss. The faster the fluid moves, the greater the losses are.

ft *abbr:* foot.

ft² *abbr:* square foot.

ft³ *abbr:* cubic foot.

ft³/bbl *abbr:* cubic feet per barrel.

ft³/d *abbr:* cubic feet per day.

ft-lb *abbr:* foot-pound

ft/min *abbr:* feet per minute.

ft³/min *abbr:* cubic feet per minute

ft/s *abbr:* feet per second.

ft³/s *abbr:* cubic feet per second.

fuel-injection nozzle *n: See nozzle.*

fuel injector *n:* a mechanical device that sprays fuel into a cylinder of an engine at the end of the compression stroke.

fuel knock *n:* a hammerlike noise produced when fuel is not burned properly in a cylinder.

fuel pump *n:* the pump that pressurizes fuel to the pressure used for injection. In a diesel engine the term is used to identify several different pumps. It is loosely used to describe the pump that transfers fuel from the main storage tank to the day tank. It is also used to describe the pump that supplies pressure to the fuel-injection pumps, although this is actually a booster-type pump.

fulcrum *n:* the support about which a lever turns.

full-gauge bit *n:* a bit that has maintained its original diameter.

full-gauge hole *n:* a wellbore drilled with a full-gauge bit. Also called a true-to-gauge hole.

full-load displacement *n:* the displacement of a mobile offshore drilling rig or ship when floating at its deepest design draft.

fungible *adj:* relating or pertaining to petroleum products with characteristics so similar that they can be mixed together, or commingled.

funnel viscosity *n:* viscosity as measured by the Marsh funnel, based on the number of seconds it takes for 1 000 cm³ of drilling fluid to flow through the funnel.

FWKO *abbr:* free-water knockout.

g *sym:* gram.

gage *n,v:* variation of *gauge*.

gal *abbr:* gallon.

gall *n:* damage to steel surfaces caused by friction and improper lubrication.

gallon *n:* a unit of measure of liquid capacity that equals 3.785 litres and has a volume of 231 in.³ A gallon of water weighs 8.34 lb at 60°F. The imperial gallon, used in Great Britain, is equal to approximately 1.2 U.S. gallons.

gal/min *abbr:* gallons per minute.

galvanic anode *n:* in cathodic protection, a sacrificial anode that produces current flow through galvanic action. See *sacrificial anode*.

galvanic cell *n:* electrolytic cell brought about by the difference in electrical potential between two dissimilar metals.

galvanic corrosion *n:* a type of corrosion that occurs when a small electric current flows from one piece of metal equipment to another. It is particularly prevalent when two dissimilar metals are present in an environment in which electricity can flow (as two dissimiliar joints of tubing in an oil or gas well).

Gamma-Gamma Density Log *n:* trade name for a density log.

gamma particle *n:* a short, highly penetrating X-ray emitted by radioactive substances during their spontaneous disintegration. The measurement of gamma particles (sometimes called gamma

rays) is the basis for a number of radioactivity well logging methods.

gamma-ray detector *n:* a device that is capable of sensing and measuring the number of gamma particles emitted by certain radioactive substances.

gamma-ray log *n:* See *radioactivity well logging*.

G&OCM *abbr:* gas- and oil-cut mud; used in drilling reports.

gang pusher *n:* the supervisor of a roustabout crew or a foreman in charge of a pipeline crew.

garbet, garbot, or **garbutt rod** *n:* a short rod on the lower end of the traveling valve of a rod pump. It is attached to the standing valve and used to pull the valve out of its seat when repairs are needed.

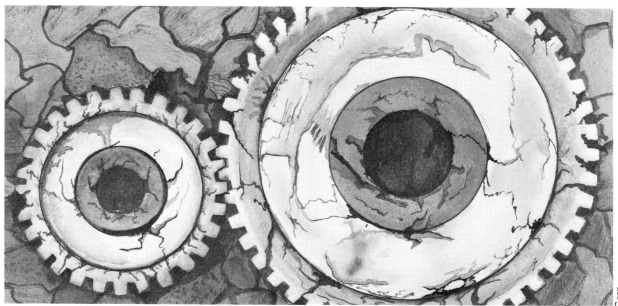

Gear

gas *n:* a compressible fluid that completely fills any container in which it is confined. Technically, a gas will not condense when it is compressed and cooled, because a gas can exist only above the critical temperature for its particular composition. Below the critical temperature, this form of matter is known as a vapor because liquid can exist and condensation can occur. Sometimes the terms *gas* and *vapor* are used interchangeably. However, the term *vapor* should be used for those streams in which condensation can occur and which originate from, or are in equilibrium with, a liquid phase.

gas anchor *n:* a tubular, perforated device attached to the bottom of a sucker-rod pump that helps to prevent gas lock. The device works on the principle that gas, being lighter than oil, rises. As well fluids enter the anchor, the gas breaks out of the fluid and exits from the anchor through perforations near the top. The remaining fluids enter the pump through a mosquito bill (a tube within the anchor), which has an opening near the bottom. In this way, all or most of the gas escapes before the fluids enter the pump.

gas cap *n:* a free-gas phase overlying an oil zone and occurring within the same producing formation as the oil. See *reservoir*.

gas-cap drive *n:* drive energy supplied naturally (as a reservoir is produced) by the expansion of gas in a cap overlying the oil in the reservoir. See *reservoir drive mechanism*.

gas-cap gas *n:* also called associated gas. See *associated gas*.

gas constant *n:* a constant number, mathematically the product of the total volume and the total pressure divided by the absolute temperature for 1 mole of any ideal gas or mixture of ideal gases at any temperature.

gas-cut mud *n:* a drilling mud that has entrained formation gas giving the mud a characteristically fluffy texture. When entrained gas is not released before the fluid returns to the well, the weight or density of the fluid column is reduced. Because a large amount of gas in mud lowers its density, gas-cut mud must be treated to reduce the chance of a blowout.

gas cutting *n:* a process in which gas becomes entrained in a liquid.

gas drilling *n:* See *air drilling*.

gas drive *n:* the use of the energy that arises from gas compressed in a reservoir to move crude oil to a wellbore. Gas drive is also used in a form of secondary recovery in which gas is injected into wells to sweep remaining oil to a producing well.

gas injection *n:* 1. injection into a reservoir of natural gas produced from the reservoir in order to maintain formation pressure and reduce the rate of decline of the original reservoir drive. Used for pressure maintenance, this method is called gas recycling or immiscible gas injection. 2. injection of a gas into a reservoir to restore reservoir pressure and to free trapped oil by miscible displacement. Used for enhanced oil recovery, this method is called miscible gas injection. A variety of gases may be used, including liquid petroleum gas, methane enriched with other light hydrocarbons, methane under high pressure, nitrogen under high pressure, and carbon dioxide. Frequently, water is also injected in alternating slugs with the gas.

gas injection well *n:* a well in which gas has been injected into an underground stratum to increase reservoir pressure.

gas input well *n:* a well into which gas is injected for the purpose of maintaining or supplementing pressure in an oil reservoir; more commonly called a gas injection well.

gasket *n:* any material (such as paper, cork, asbestos, or rubber) used to seal two essentially stationary surfaces.

gas lift *n:* the process of raising or lifting fluid from a well by injecting gas down the well through tubing or through the tubing-casing annulus. Injected gas aerates the fluid to make it exert less pressure than the formation does; consequently, the higher formation pressure forces the fluid out of the wellbore. Gas may be injected continuously or intermittently, depending on the producing characteristics of the well and the arrangement of the gas-lift equipment.

gas-lift mandrel *n:* a device installed in the tubing string of a gas-lift well onto which or into which a gas-lift valve is fitted. There are two common types of mandrels. In the conventional gas-lift mandrel, the gas-lift valve is installed as the tubing is placed in the well. Thus, to replace or repair the valve, the tubing string must be pulled. In the sidepocket mandrel, however, the valve is installed and removed by wireline while the mandrel is still in the well, eliminating the need to pull the tubing to repair or replace the valve.

gas-lift valve *n:* a device installed on a gas-lift mandrel, which in turn is put on the tubing string of a gas-lift well. Tubing and casing pressures cause the valve to open and close, thus allowing gas to be injected into the fluid in the tubing to cause the fluid to rise to the surface.

gas liquids n pl: See *liquefied petroleum gas.*

gas lock n: a condition sometimes encountered in a pumping well when dissolved gas, released from solution during the upstroke of the plunger, appears as free gas between the valves. If the gas pressure is sufficient, the standing valve is locked shut, and consequently no fluid enters the tubing.

gas-oil contact n: the point or plane in a reservoir at which the bottom of a gas sand is in contact with the top of an oil sand.

gas-oil ratio n: a measure of the volume of gas produced with oil, expressed in cubic feet per barrel or cubic metres per tonne.

gasoline n: a volatile, flammable liquid hydrocarbon refined from crude oils and used universally as a fuel for internal-combustion, spark-ignition engines.

gasoline plant n: also called a natural gas processing plant, a term that is preferred because it distinguishes the plant from a unit that makes gasoline within an oil refinery. See *natural gas processing plant.*

gas pipeline n: a transmission system for natural gas or other gaseous material. The total system is comprised of required pipes and compressors needed to maintain the flowing pressure of the system.

gas processing n: the separation of constituents from natural gas for the purpose of making salable products and also for treating the residue gas to meet required specifications.

gas processing plant n: also called natural gas processing plant. See *natural gas processing plant.*

gas regulator n: an automatically operated valve that, by opening and closing in response to pressure, permits more or less gas to flow through a pipeline and thus controls the pressure.

gas reservoir n: a geological formation containing a single gaseous phase. When produced, the surface equipment may or may not contain condensed liquid, depending on the temperature, pressure, and composition of the single reservoir phase.

gas sand n: a stratum of sand or porous sandstone from which natural gas is obtained.

gasser n: a well that produces natural gas.

gas show n: the gas that appears in drilling fluid returns, indicating the presence of a gas zone.

gas transmission system n: the central or trunk pipeline system by means of which dry natural gas is transported from field gathering stations or processing plants to the industrial or domestic fuel market. Well pressure is supplemented at intervals along the transmission line by compressors to maintain a flow strong enough to move the gas to its destination.

gas turbine n: an engine in which gas, under pressure or formed by combustion, is directed against a series of turbine blades. The energy in the expanding gas is converted into rotary motion.

gas well n: a well that primarily produces gas. Legal definitions vary among the states.

gate n: a shutter device in a logging tool that allows an exposure-time measurement, that is, 400-600 microseconds.

gate valve n: a type of valve. See *valve.*

gathering line n: a pipeline, usually of small diameter, used in gathering crude oil or gas from the field to a main pipeline.

gathering system n: the pipelines and other equipment needed to transport oil, gas, or both from wells to a central point – the gathering station – where there is the accessory equipment required to deliver a clean and salable product to the market or to another pipeline. An oil gathering system includes oil and gas separators, emulsion treaters, gathering tanks, and similar equipment. A gas gathering system includes regulators, compressors, dehydrators, and associated equipment.

gauge n: 1. the diameter of a bit or the hole drilled by the bit. 2. a device used to measure some physical property (such as a pressure gauge). v: to measure size, volume, or other measurable property.

gauge cutters n: the teeth or tungsten carbide inserts in the outermost row on the cones of a bit, so called because they cut the outside edge of the hole and determine the hole's gauge or size. Also called heel teeth.

gauge height n: the distance from the gauge, or reference, point to the bottom of a tank. On marine vessels, the measurement must be made when the vessel is on even keel.

gauge point n: the point at which the tape is lowered and gauge is read on a tank, usually at the rim of the ullage hatch, manway, or expansion dome.

gauge pressure n: the amount of pressure exerted on the interior walls of a vessel by the fluid contained in it (as indicated by a pressure gauge); it is expressed in psig (pounds per square inch gauge) or in kilopascals. Gauge pressure plus atmospheric pressure equals absolute pressure. See *absolute pressure.*

gauger n: a pipeline representative for the sale or transfer of crude oil from the producer to the pipeline. He samples and tests the crude oil to determine quantity and quality, and uses a calibrated, flexible-steel tape with a plumb bob at the end to measure the oil in the tank.

gauge-row patterns n: the configuration or shape of the teeth in the outermost row of teeth on a bit.

gauge surface n: the outside surfaces on the outermost rows of teeth on a bit; they determine the diameter or gauge of the hole to be drilled.

gauging hatch n: the opening in a tank or other vessel through which measuring and sampling are performed.

gauging tables n pl: tables prepared by engineers to show the calculated number of barrels or cubic metres for any given depth of liquid in a tank. They are sometimes called strapping tables.

gauging tape n: a metal tape used to measure the depth of liquid in a tank.

GC abbr: gas-cut; used in drilling reports.

GCM abbr: gas-cut mud; used in drilling reports.

GDC survey n: density log in which a gamma ray log, a Densilog, and a caliper log are recorded simultaneously.

gear n: a toothed wheel made to mesh with another toothed wheel.

gel n: a semisolid, jellylike state assumed by some colloidal dispersions at rest. When agitated, the gel converts to a fluid state. Also a nickname for bentonite. v: to take the form of a gel; to set.

gel cement n: cement or cement slurry that has been modified by the addition of bentonite.

gel strength n: a measure of the ability of a colloidal dispersion to develop and retain a gel form, based on its resistance to shear. The gel strength, or shear strength, of a drilling mud determines its ability to hold solids in suspension. Sometimes bentonite and other colloidal clays are added to drilling fluid to increase its gel strength. See *gel*.

generator n: a machine that changes mechanical energy into electrical energy.

geologic time scale n: the long periods of time dealt with and identified by geology. Geologic time is divided into eras (Cenozoic, Mesozoic, Cambrian, and Precambrian), which are subdivided into periods and epochs. When the age of a type of rock is determined, it is assigned a place in the scale and thereafter referred to as, for example, Mesozoic rock of the Triassic period.

geologist n: a scientist who gathers and interprets data pertaining to the strata of the earth's crust.

Geolograph n: trade name for a patented device that automatically records the rate of penetration and depth during drilling.

geology n: the science that relates to the study of the structure, origin, history, and development of the earth and its inhabitants as revealed in the study of rocks, formations, and fossils.

geophone n: an instrument that detects vibrations passing through the earth's crust, used in conjunction with seismography. Geophones are often called jugs. See *seismograph*.

geopressure n: abnormally high pressure exerted by some subsurface formations. The deeper the formation lies, the higher is the pressure it exerts on a wellbore drilled into it.

geosyncline n: a part of the earth's surface that sank over a long period of time, forming a trough hundreds of miles long and tens of miles wide. Thousands of feet of sedimentary and volcanic rock were formed in it over millions of years.

geothermal gradient n: the increase in the temperature of the earth with increasing depth. It averages about 1°F per 60 feet but may be considerably higher or lower.

geothermal reservoir n: 1. a subsurface layer of rock containing steam or hot water that is trapped in the layer by overlying impermeable rock. 2. a subsurface layer of rock that is hot but contains little or no water. Geothermal reservoirs are a potential source of energy.

geronimo n: See *safety slide*.

gilsonite n: a naturally occurring solid hydrocarbon belonging to the asphalt group; it is lustrous black with brown streaks. A granular form of gilsonite is sometimes used as a cement additive to prevent lost circulation.

gimbal n: a mechanical frame that permits an object mounted in it to remain in a stationary or near stationary position regardless of movement of the frame. Gimbals are often used offshore to counteract undesirable wave motion.

gin pole n: a pole (usually single) supported with guy wires and used with block and tackle to hoist equipment. On a drilling rig the gin pole is typically secured to the mast or derrick above the monkeyboard.

gin pole truck n: a truck equipped with hoisting equipment and a pole or arrangement of poles for use in lifting heavy machinery.

girt *n:* one of the horizontal braces between the legs of a derrick.

GL *abbr:* ground level; used in drilling reports.

gland *n:* a device used to form a seal around a reciprocating or rotating rod (as in a pump) to prevent fluid leakage; specifically, the movable part of a stuffing box by which the packing is compressed. See *stuffing box.*

gland-packing nut *n:* a threaded device, the sides of which are arranged so that a wrench can be fitted onto them; used to retain the gland packing in place around a rod. See *gland packing.*

globe valve *n:* See *valve.*

glow plug *n:* a small electric heating element placed inside a diesel engine cylinder to preheat the air and to make starting easier.

glycol *n:* a group of compounds used to dehydrate gaseous or liquid hydrocarbons or to inhibit the formation of hydrates. Commonly used glycols are ethylene glycol, diethylene glycol, and triethylene glycol.

glycol dehydrator *n:* a processing unit used to remove all or most of the water from gas. Usually a glycol unit includes a tower, in which the wet gas is put into contact with glycol to remove the water, and a reboiler, which heats the wet glycol to remove the water from it so that it can be recycled.

go-devil *n:* 1. a device that is inserted into a pipeline for the purpose of cleaning; a line scraper. Also called a pig. 2. a device that is lowered into the borehole of a well for various purposes such as enclosing surveying instruments, detonating instruments, and the like. *v:* to drop or pump a device down the borehole, usually through drill pipe or tubing.

go in the hole *n:* to lower the drill stem into the wellbore.

gone to water *adj:* pertaining to a well in which production of oil has decreased and production of water has increased (e.g., "the well has gone to water").

gooseneck *n:* the curved connection between the rotary hose and the swivel. See *swivel.*

GOR *abbr:* gas-oil ratio.

governor *n:* any device that limits or controls the speed of an engine.

gpm *abbr:* gallons per minute.

gr *abbr:* gray; used in drilling reports.

graben *n:* a block of the earth's crust that has slid downward between two faults; the opposite of a horst.

graded, or mixed, string *n:* a casing string made up of several weights or grades of casing, and designed to take into account well depth, expected pressures, and weight of the fluid in the well.

gram *n:* a unit of metric measure of mass and weight equal to 1/1,000 kg and nearly equal to 1 cm^3 of water at its maximum density.

gram molecular weight *n:* See *molecular weight.*

granite *n:* an igneous rock composed primarily of feldspar, quartz, and mica. It usually does not contain petroleum.

grass roots refinery *n:* a refinery built from the ground up, as opposed to one to which an addition or a modification has been made.

gravel pack *n:* a mass of very fine gravel placed around a slotted liner in a well. See *liner.*

gravel-pack *v:* to place a slotted or perforated liner in a well and surround it with small-sized gravel. See *gravel packing.*

gravel packing *n:* a method of well completion in which a slotted or perforated liner is placed in the well and surrounded by small-sized gravel. The well is sometimes enlarged by underreaming at the point where the gravel is packed. The mass of gravel excludes sand from intruding in the well but allows continued rapid production.

gravel-pack packer *n:* a packer used for the well completion method of gravel packing. See *gravel packing.*

graveyard tour *n:* (pronounced "tower") the shift of duty on a drilling rig that starts at midnight; sometimes called the morning tour.

gravimeter *n:* an instrument used to detect and measure minute differences in the earth's gravitational pull at different locations to obtain data about subsurface formations.

gravitometer *n:* a device for measuring and recording the specific gravity of a gas or liquid passing a point of measurement.

gravity *n:* the attraction exerted by the earth's mass on objects at its surface; the weight of a body. See *API gravity* and *specific gravity.*

gravity drainage *n:* the movement of the oil in a reservoir toward a wellbore, resulting from the force of gravity. In the absence of water drive or effective gas drive, gravity drainage is an important source of energy to produce oil. It is also called segregation drive.

grease fitting *n:* a device on a machine, designed to accept the hose of a grease gun so that grease can be added to the part in need of lubrication.

greensand *n:* a sand that contains considerable quantities of glauconite, a greenish mineral composed of potassium, iron, and silicate, which gives the sand its color and name.

grief stem *n:* (obsolete) kelly; kelly joint.

grind out *v:* to test for the presence of water in oil by use of a centrifuge.

grn *abbr:* green; used in drilling reports.

gross observed volume *n:* the total volume of all petroleum liquids and sediment and water, excluding free water, at observed temperature and pressure.

gross production *n:* the total production of oil from a well or lease during a specified period of time.

gross standard volume *n:* the total volume of all petroleum liquids and sediment and water, excluding free water, corrected by the appropriate temperature correction factor (C_{tl}) for the observed temperature and API gravity, relative density, or density to a standard temperature such as 60°F or 15°C, and also corrected by the applicable pressure correction factor (C_{pl}) and meter factor.

gross tonnage *n:* the interior capacity of a ship or a mobile offshore drilling unit. The capacity is expressed in tons, although the actual measurement is in volume, 1 ton being equivalent to 100 cubic feet of volume. This rule holds for measuring ship capacity for U.S.A. maritime purposes. All principal maritime governments have their own rules describing how tonnage is to be measured.

ground anchor *n:* also called a deadman. See *deadman*.

ground bed *n:* in cathodic protection, an interconnected group of impressed-current anodes that absorbs the damage caused by generated electric current.

ground block *n:* a wireline sheave, or pulley, that is fastened to the ground anchor and that changes a horizontal pull on a wireline to a vertical pull (as in swabbing with a derrick over a well). See *block*.

grout *v:* to force sealing material into a soil, sand, or rock formation to stabilize it. *n:* the sealing material used in grouting.

guard *n:* a metal shield placed around moving parts of machinery to lessen or avoid the chance of injury to personnel. In the oil field, guards are used on equipment such as belts, power transmission chains, drums, flywheels, and drive shafts.

guard-electrode log *n:* a focused system designed to measure the true formation resistivity in boreholes filled with salty mud. Current is forced by guard electrodes to flow into the formation. A drop in potential difference of controlled current between the electrodes is created by the flow of the current through the surrounding formation to a remote current-return electrode. This potential difference serves as a measure of the resistivity of the formation.

guardrail *n:* a railing for guarding against danger or trespass. On a drilling or workover rig, for example, guardrails are used on the rig floor to prevent persons from falling; guardrails are also installed on the mud pits and other high areas where there is any danger of falling.

guide base *n:* See *temporary guide base* and *permanent guide base*.

guide fossil *n:* the petrified remains of plants or animals, useful for correlation and age determination of the rock in which they were found.

guidelines *n pl:* lines, usually four, attached to the temporary guide base and permanent guide base, that help to position equipment (such as blowout preventers) accurately on the seafloor when a well is drilled offshore.

guide shoe *n:* a short, heavy, cylindrical section of steel, filled with concrete and rounded at the bottom, which is placed at the end of the casing string. It prevents the casing from snagging on irregularities in the borehole as it is lowered. A passage through the center of the shoe allows drilling fluid to pass up into the casing while it is being lowered and allows cement to pass out during cementing operations. Also called casing shoe.

gumbo *n:* any relatively sticky formation (such as clay) encountered in drilling.

gun barrel *n:* a settling tank used to separate oil and water in the field. After emulsified oil is heated and treated with chemicals, it is pumped into the gun barrel, where the water settles out and is drawn off, and the clean oil flows out to storage. Gun barrels have largely been replaced by unified heater-treater equipment but are still common, especially in older fields.

Gunite *n:* trade name for a cement-sand mixture used to seal pipe against air, moisture, and corrosion damage.

gunk squeeze *n:* a bentonite and diesel-oil mixture that is pumped down the drill pipe to mix with drilling mud being pumped down the annulus. The stiff, puttylike material is squeezed into lost circulation zones to isolate them from the wellbore.

gun-perforate *v:* to create holes in casing and cement set through a productive formation. A common method of completing a well is to set casing through the oil-bearing formation and cement it. A perforating gun is then lowered into the hole and fired to detonate high-powered jets or shoot steel projectiles (bullets) through the casing and cement and into the pay zone. The formation fluids flow out of the reservoir through the perforations and into the wellbore. See *jet-perforate* and *perforating gun*.

gusher *n:* an oilwell that has come in with such great pressure that the oil jets out of the well like a geyser. In reality, a gusher is a blowout and is extremely wasteful of reservoir fluids and drive energy. In the early days of the oil industry, gushers were common and many times were the only indications that a large reservoir of oil and gas had been struck. See *blowout*.

guying system *n:* the system of guy lines and anchors used to brace a rig. See *guy line*.

guy line *n:* a wireline attached to a mast or derrick to stabilize it. See *wind guy line* and *load guy line*.

guy line anchor *n:* a buried weight or anchor to which a guy line is attached. See *deadman*.

gyp *n:* slang for gypsum. See *gypsum*.

gypsum *n:* a naturally occurring crystalline form of calcium sulfate where each molecule of calcium sulfate is combined with two molecules of water. See *calcium sulfate* and *anhydrite*.

gyrate *v:* to move spirally about an axis or to revolve around a central point.

gyroscopic surveying instrument *n:* a device used to determine direction and angle at which a wellbore is drifting off the vertical. Unlike a magnetic surveying instrument, a gyroscopic instrument obtains direction well and is not affected by magnetic irregularities that may be caused by casing or other ferrous metals. See *directional survey* and *directional drilling*.

H

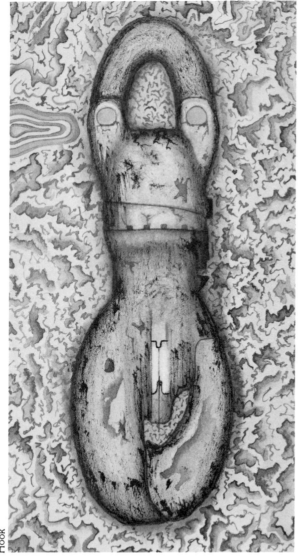

Hook

h *abbr:* hour.

H *sym:* henry.

half-cell *n:* a single electrode immersed in an electrolyte for the purpose of measuring metal-to-electrolyte potentials and, therefore, the corrosion tendency of a particular system.

half-life *n:* the amount of time needed for half of a quantity of radioactive substance to decay or transmute into a nonradioactive substance. Half-lives range from fractions of seconds to millions of years.

hammer drill *n:* a drilling tool that, when placed in the drill stem just above a roller cone bit, delivers high-frequency percussion blows to the rotating bit. Hammer drilling combines the basic features of rotary and cable-tool drilling (i.e., bit rotation and percussion).

hammer-drill *v:* to use a hammer drill. See *hammer drill.*

hammer test *n:* a method of locating corroded sections of pipe by striking the pipe with a hammer. When struck, a corroded section resounds differently than a noncorroded section.

hand *n:* a worker in the oil industry, especially one in the field.

handling-tight coupling *n:* a coupling screwed onto casing tight enough so that a wrench must be used to remove the coupling.

handrail *n:* a railing or pipe along a passageway or stair that serves as a support or a guard.

hanger plug *n:* a device placed or hung in the casing below the blowout preventer stack to form a pressure-tight seal. Pressure is then applied to the blowout preventer stack in order to test it for leaks.

hangline *n:* a single length of wire rope attached to the crown block by which the traveling block is suspended when not in use. Also called a hang-off line.

hang-off line *n:* also called hangline. See *hangline.*

hang rods *v:* to suspend sucker rods in a derrick or mast on rod hangers rather than to place them horizontally on a rack.

hard banding *n:* a special wear-resistant material often applied to tool joints in order to prevent abrasive wear to the area when the pipe is being rotated downhole.

hardfacing *n:* an extremely hard material, usually crushed tungsten carbide, that is applied to the outside surfaces of tool joints, drill collars, stabilizers, and other rotary drilling tools to minimize wear when they are in contact with the wall of the hole.

hard hat *n:* a metal or hard-plastic helmet worn by oil field workers to minimize the danger of being injured by falling objects.

hard water *n:* water that contains dissolved compounds of calcium, magnesium, or both. Compare *soft water.*

hatch *n:* 1. an opening in the roof of a tank through which a gauging line may be lowered to measure its contents. 2. the opening from the deck into the cargo space of ships.

hayrack *n:* (obsolete) a rack used to hold pipe on a derrick; a fingerboard.

hayrake *n:* also called hayrack. See *hayrack.*

hay section *n:* a section of a heater or a heater-treater that is filled with fibrous material through which oil and water emulsions are filtered.

hazard *n:* an object or a condition, related to equipment, site, or environment, that presents or causes a risk of accident or fire.

HCN *form:* hydrogen cyanide.

head *n:* 1. the height of a column of liquid required to produce a specific pressure. See *hydraulic head.* 2. for centrifugal pumps, the velocity of flowing fluid converted into pressure expressed in feet or metres of flowing fluid. Also called velocity head. 3. that part of a machine (such as a pump or an engine) that is on the end of the cylinder opposite the crankshaft.

headache post *n:* the post on cable-tool rigs that supports the end of the walking beam when the rig is not operating.

headgate *n:* the gate valve nearest the pump or compressor on oil or gas lines.

heading *n:* an intermittent flow of oil from a well.

head well puller *n:* crew chief.

heater *n:* a container or vessel enclosing an arrangement of tubes and a firebox in which an emulsion is heated before further treating.

heater-treater *n:* a vessel that heats an emulsion and removes water and gas from the oil to raise it to a quality acceptable for pipeline transmission. A heater-treater is a combination of heater, free-water knockout, and oil and gas separator.

heat exchanger *n:* See *exchanger.*

heating coils *n pl:* (marine) a system of piping in tank bottoms, in which steam is carried as required to heat high pour-point liquid cargoes to pumpable viscosity level.

heating medium *n:* a material, whether flowing or static, used to transport heat from a primary source such as combustion of fuel to another material. Heating oil and steam are examples of heating mediums. Also called heat medium.

heating value *n:* the amount of heat developed by the complete combustion of a unit quantity of a material. Also called heat of combustion.

heat of combustion *n:* also called heating value. See *heating value.*

heave *n:* the vertical motion of a ship or a floating offshore drilling rig.

heave compensator *n:* a device that moves with the heave of a floating offshore drilling rig to prevent the bit from being lifted off the bottom of the hole and then dropped back down (i.e., to maintain constant weight on the bit). It is used with devices such as bumper subs. See *motion compensator.*

heavy ends *n:* 1. the parts of a hydrocarbon mixture that have the highest boiling point and the highest viscosity (such as fuel oils and waxes). 2. hexanes and heptanes in a natural gas stream.

heavyweight additive *n:* a substance or material added to cement to make it dense enough for use in high-pressure zones. Sand, barite, and hematite are some of the substances used as heavyweight additives.

heavyweight drill pipe *n:* drill pipe having thicker walls and longer tool joints than usual and also an integral wear pad in the middle. Several joints of this pipe may be placed in the drill stem between drill collars and regular drill pipe to reduce the chances of drill pipe fatigue or failure.

heel *n:* the inclination of a ship or a floating offshore drilling rig to one side, caused by wind, waves, or shifting weights on board.

heel teeth *n:* also called gauge cutters. See *gauge cutters.*

helical *adj:* spiral-shaped in design.

helical spooling *n:* spooling of wire rope onto a drum on which the grooves take the form of a helix – that is, like the threads on a pipe end.

helium unscrambler *n:* an electronic device that slows down the voice of a diver using helium as part of his breathing mixture so that the surface crew may understand him.

helmet *n:* a protective enclosure for a diver's entire head. It is part of his life-support system and also contains a communications system.

henry *n:* the unit of self-inductance or mutual inductance in the metric system. Its symbol is H.

heptane *n:* a saturated hydrocarbon of the paraffin series; one of the heavy ends in a hydrocarbon mixture.

hertz *n:* a unit in the metric system used to measure frequency in cycles per second. Its symbol is Hz.

hexane *n:* a saturated hydrocarbon of the paraffin series; one of the heavy ends in a hydrocarbon mixture.

hexanes-plus (or **heptanes-plus**) *n pl:* the portion of a hydrocarbon fluid mixture or the last component of a hydrocarbon analysis that contains the hexanes (or heptanes) and all hydrocarbons heavier than the hexanes (or heptanes).

HGOR *abbr:* high gas-oil ratio; used in drilling reports.

high drum drive *n:* the drive for the drawworks drum used when hoisting loads are light.

high-pressure nervous syndrome *n:* a term used to describe symptoms caused by high partial pressures of helium.

high-pressure squeeze cementing *n:* the forcing of cement slurry into a well at the points to be sealed with a final pressure equal to or greater than the formation breakdown pressure. See *squeeze cementing.*

high-purity water *n:* water that has little or no ionic content and is therefore a poor conductor of electricity.

HOCM *abbr:* heavily oil-cut mud; used in drilling reports.

hogging *n:* the distortion of the hull of an offshore drilling rig when the bow and the stern are lower than the middle, caused by wave action or unbalanced or heavy loads; the opposite of sagging.

hoist *n:* 1. an arrangement of pulleys and wire rope or chain used for lifting heavy objects; a winch or similar device. 2. the drawworks. See *drawworks. v:* to raise or lift.

hoisting cable *n:* the cable that supports drill pipe, swivel, hook, and traveling block on a rotary drilling rig.

hoisting components *n:* drawworks, drilling line, and traveling and crown blocks. Auxiliary hoisting components include catheads, catshaft, and air hoist.

hoisting drum *n:* the large flanged spool in the drawworks on which the hoisting cable is wound. See *drawworks.*

hoisting plug *n:* also called a lifting sub or a lifting nipple. See *lifting nipple.*

hold-down *n:* a mechanical arrangement that prevents the upward movement of certain pieces of equipment installed in a well. A sucker rod pump may use a mechanical hold-down for attachment to a seating nipple.

hole *n:* 1. in drilling operations, the wellbore or borehole. See *wellbore* and *borehole.* 2. an opening that is made purposely or accidentally in any solid substance.

hole opener *n:* a device used to enlarge the size of an existing borehole, having teeth arranged on its outside circumference to cut the formation as it rotates.

holiday *n:* a gap or void in coating on a pipeline or in paint on a metal surface.

holiday detector *n:* an electrical device used to locate a weak place, or holiday, in coatings on pipelines and equipment.

hollow carrier gun *n:* a perforating gun consisting of a hollow, cylindrical metal tube into which are loaded shaped charges or bullets. Upon detonation, debris caused by the exploding charges falls into the carrier to be retrieved with the reusable gun.

homocline *n:* a series of beds dipping in the same direction.

honeycomb formation *n:* a stratum of rock that contains large void spaces; a cavernous or vugular formation.

hook *n:* a large, hook-shaped device from which the swivel is suspended. It is designed to carry maximum loads ranging from 100 to 650 tons (90 to 590 tonnes) and turns on bearings in its supporting housing. A strong spring within the assembly cushions the weight of a stand (90 feet or about 27 metres) of drill pipe, thus permitting the pipe to be made up and broken out with less damage to the tool joint threads. Smaller hooks without the spring are used for handling tubing and sucker rods.

hook load capacity *n:* the nominal rated load capacity of a portable hoist and mast arrangement, usually calculated by an API formula.

hook-wall packer *n*: a packer equipped with friction blocks or drag springs and slips and designed so that rotation of the pipe unlatches the slips. The friction springs prevent the slips and the hook from turning with the pipe and assist in advancing the slips up a tapered sleeve to engage the wall of the outside pipe as weight is put on the packer. Also called a wall-hook packer. See *packer*.

hopper *n*: a large funnel- or cone-shaped device into which dry components (such as powdered clay or cement) can be poured in order to uniformly mix the components with water or other liquids. The liquid is injected through a nozzle at the bottom of the hopper. The resulting mixture of dry material and liquid may be drilling mud to be used as the circulating fluid in a rotary drilling operation, or it may be cement slurry to be used in bonding casing to the borehole.

horse head *n*: the curved section of the walking beam of a beam pumping unit, which is located on the oilwell end and from which the bridle is suspended.

horsepower *n*: a unit of measure of work done by a machine. One horsepower equals 33,000 foot-pounds per minute.

horst *n*: a block of the earth's crust that has been raised up between two faults; the opposite of a graben.

hot carbonate process *n*: a process for removing the bulk of acid gases from a gas stream by contacting the stream with a water solution of potassium carbonate at a temperature in the range of 220°F to 240°F (104°C to 116°C).

hot oil *n*: 1. absorption or other oil used as a heating medium. 2. oil produced in violation of state regulations or transported interstate in violation of federal regulations.

hot-oil treatment *n*: the treatment of a producing well with heated oil to melt accumulated paraffin in the tubing and the annulus.

hot spot *n*: an abnormally hot place on a casing coupling when a joint is being made up. It usually indicates worn threads on the pipe and in the coupling.

housing *n*: something that covers or protects, like the casing for a mechanical part.

hp *abbr*: horsepower.

hp-h *abbr*: horsepower-hour.

HPNS *abbr*: high-pressure nervous syndrome.

H$_2$S *form*: hydrogen sulfide.

H$_2$SO$_3$ *form*: sulfurous acid.

H$_2$SO$_4$ *form*: sulfuric acid.

huff-and-puff injection *n*: (slang) cyclic steam injection.

hull *n*: the framework of a vessel including all decks, plating, and columns, but excluding machinery.

hunting *n*: a surge of engine speed to a higher number of revolutions per minute (rpm), followed by a drop to normal speed without manual movement of the throttle. It is often caused by a faulty or improperly adjusted governor.

Hydrafrac *n*: the copyrighted name of a method of hydraulic fracturing for increasing productivity.

hydrate *n*: a hydrocarbon and water compound that is formed under reduced temperature and pressure in gathering, compression, and transmission facilities for gas. Hydrates often accumulate in troublesome amounts and impede fluid flow. They resemble snow or ice. *v*: to enlarge by taking water on or in.

hydrated lime *n*: calcium hydroxide; a dry powder obtained by treating quicklime with enough water to satisfy its chemical affinity for water. Its chemical formula is $Ca(OH)_2$.

hydration *n*: reaction of cement with water. The powdered cement gradually sets to a solid as hydration continues.

hydraulic *adj*: 1. of or relating to water or other liquid in motion. 2. operated, moved, or effected by water or liquid.

hydraulic brake *n*: also called hydrodynamic brake or Hydromatic® brake. See *hydrodynamic brake*.

hydraulic control pod *n*: a device used on offshore drilling rigs to provide a way to actuate and control subsea blowout preventers from the rig. Hydraulic lines from the rig enter the pods, through which fluid is sent toward the preventer. Usually two pods, painted different colors, are used, each to safeguard and back up the other.

hydraulic coupling *n*: a fluid connection between a prime mover and the machine it drives; it uses the action of liquid moving against blades to drive the machine.

hydraulic fluid *n*: a liquid of low viscosity (such as light oil) that is used in systems actuated by liquid (such as the brake system in a modern passenger car).

hydraulic fracturing *n:* an operation in which a specially blended liquid is pumped down a well and into a formation under pressure high enough to cause the formation to crack open. The resulting cracks or fractures serve as passages through which oil can flow into the wellbore. See *formation fracturing*.

hydraulic head *n:* the force exerted by a column of liquid expressed by the height of the liquid above the point at which the pressure is measured. Although *head* refers to distance or height, it is used to express pressure, since the force of the liquid column is directly proportional to its height. Also called head or hydrostatic head. Compare *hydrostatic pressure*.

hydraulic jar *n:* also called mechanical jar. See *mechanical jar*.

hydraulic pump *n:* a device that lifts oil from wells without the use of sucker rods. See *hydraulic pumping*.

hydraulic pumping *n:* a method of pumping oil from wells by using a downhole pump without sucker rods. Subsurface hydraulic pumps consist of two reciprocating pumps coupled and placed in the well. One pump functions as an engine and drives the other pump (the production pump). Surface power is supplied from a standard engine-driven pump. The downhole engine is operated by clean crude oil under pressure (power oil) that is drawn from a power-oil or settling tank by a triplex plunger pump. If a single string of tubing is used, power oil is pumped down the tubing string to the pump, which is seated in the string, and a mixture of power oil and produced fluid is returned through the casing-tubing annulus. If two parallel strings are used, one supplies the power oil to the pump while the other returns the exhaust and the produced oil to the surface. The hydraulic pump may be used to pump several wells from a central source and has been used to lift oil from depths of more than 10,000 feet (3 048 m).

hydraulic ram *n:* a cylinder and piston device that uses hydraulic pressure for pushing, lifting, or pulling. It is commonly used to raise portable masts from a horizontal to a vertical position, for leveling a production rig at an uneven location, or for closing a blowout preventer against pressure.

hydraulics *n:* the branch of science that deals with practical applications of water or other liquid in motion.

hydraulic torque wrench *n:* a hydraulically powered device that can break out or make up tool joints and assure accurate torque. It is fitted with a repeater gauge so that the driller can monitor tool joints as they go downhole, doubly assuring that all have the correct torque. Sometimes called an iron roughneck.

Hydril *n:* the registered trademark of a prominent manufacturer of oil field equipment, especially the annular blowout preventer.

hydrocarbons *n pl:* organic compounds of hydrogen and carbon, whose densities, boiling points, and freezing points increase as their molecular weights increase. Although composed of only two elements, hydrocarbons exist in a variety of compounds because of the strong affinity of the carbon atom for other atoms and for itself. The smallest molecules of hydrocarbons are gaseous; the largest are solids. Petroleum is a mixture of many different hydrocarbons.

hydrochloric acid *n:* an acid compound commonly used to acidize carbonate rocks; prepared by mixing hydrogen chloride gas in water. Also known as muriatic acid. Its chemical formula is HCl.

hydrocracking *n:* cracking in the presence of low-pressure hydrogen, consuming a net amount of hydrogen in the process.

hydrocyclone *n:* a cone-shaped separator for separating various sizes of particles and liquid by centrifugal force. See *desander* and *desilter*.

hydrodynamic brake *n:* a device mounted on the end of the drawworks shaft of a drilling rig. The hydrodynamic brake serves as an auxiliary to the mechanical brake when pipe is lowered into the well. The braking effect of a hydrodynamic brake is achieved by means of an impeller turning in a housing filled with water. Sometimes called hydraulic brake or Hydromatic® (a manufacturer's term) brake.

hydrofluoric-hydrochloric acid *n:* a mixture of acids used for removal of mud from the wellbore. See *mud acid*.

hydroforming *n:* a process of petroleum refining in which straight-run, cracked, or mixed naphthas are passed over a solid catalyst at elevated temperatures and moderate pressures in the presence of added hydrogen or hydrogen-containing gases. The main chemical reactions are dehydrogenation and aromatization of the nonaromatic constituents of the naphtha to form either high-octane motor fuel or high-grade aviaton gasoline high in aromatic hydrocarbons such as toluene, xylenes, and so forth. Sulfur contained in the naphtha is 90% removed.

hydrogen cyanide *n:* an extremely poisonous compound of hydrogen, carbon, and nitrogen (HCN), with a boiling point of 79°F (26°C) and having the odor of bitter almonds. It is water-soluble in all proportions. Also called hydrocyanic acid, prussic acid, or formonitrile.

hydrogen embrittlement *n:* also called acid brittleness. See *acid brittleness*.

hydrogen patch probe *n:* an instrument that, when attached to the exterior of a vessel that has been corroded by hydrogen sulfide, senses the hydrogen content in the steel and records the rate of corrosion.

hydrogen richness *n:* the amount of a formation's hydrogen content.

hydrogen sulfide *n:* a flammable, colorless gaseous compound of hydrogen and sulfur (H_2S) with the odor of rotten eggs. Commonly found in petroleum, it causes the foul smell of petroleum fractions. It is extremely corrosive and poisonous, causing damage to skin, eyes, breathing passages, and lungs and attacking and paralyzing the nervous system, particularly that part controlling the lungs and heart. Also called hepatic gas or sulfureted hydrogen.

Hydromatic® brake *n:* trade name for a type of hydrodynamic brake.

hydrometer *n:* an instrument with a graduated stem, used to determine the gravity of liquids. The liquid to be measured is placed in a cylinder, and the hydrometer dropped into it. It floats at a certain level in the liquid (high if the liquid is light, low if it is heavy), and the stem markings indicate the gravity of the liquid.

hydrophilic *adj:* tending to adsorb water.

hydrophobic *adj:* tending to repel water.

hydrostatic pressure *n:* the force exerted by a body of fluid at rest; it increases directly with the density and the depth of the fluid and is expressed in psi or kPa. The hydrostatic pressure of fresh water is 0.433 psi per foot of depth (9.792 kPa/m). In drilling, the term refers to the pressure exerted by the drilling fluid in the wellbore. In a water-drive field, the term refers to the pressure that may furnish the primary energy for production.

hydro-test *v:* to apply hydraulic pressure to check for leaks in tubing or tubing couplings, usually as the tubing is being run into the well. If water leaks from any place in the tubing, the joint of tubing, the coupling, or both are replaced.

hypercapnia *n:* excessive amount of carbon dioxide in the blood, often resulting from an excessive carbon dioxide partial pressure in a diver's breathing supply. Also called carbon dioxide excess.

hypothermia *n:* reduced body temperature caused by overexposure to chilling temperatures.

Hz *sym:* hertz.

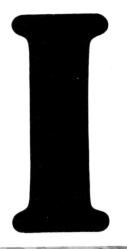

IADC *abbr:* International Association of Drilling Contractors, formerly the American Association of Oilwell Drilling Contractors (AAODC).

ICC *abbr:* Interstate Commerce Commission.

ID *abbr:* inside diameter.

idiot stick *n:* (slang) a shovel.

idle *v:* to operate an engine without applying a load to it.

idler *n:* a pulley or sprocket used with belt or chain drives on machinery to maintain desired tension on the belt or chain.

I-ES *abbr:* induction-electric survey.

ig *abbr:* igneous; used in drilling reports.

igneous rock *n:* a rock mass formed by the solidification of material poured (when molten) into the earth's crust or onto its surface. Granite is an igneous rock.

ignition quality *n:* the ability of a fuel to ignite when it is injected into the compressed-air charge in a diesel engine cylinder. It is measured by an index called the cetane number.

ignorant end *n:* (slang) the heavier end of any device (such as a length of pipe or a wrench).

ilmenite *n:* an iron-black mineral of composition $FeTiO_3$ or $FeO \cdot TiO_2$, with a specific gravity of about 4.67, sometimes used for increasing the density of oilwell cement slurries.

IMCO *abbr:* Intergovernmental Maritime Consultative Organization.

immiscible *adj:* not capable of mixing (as oil and water).

IMP *abbr:* Instituto Mexicano del Petróleo (Mexican Petroleum Institute).

impeller *n:* a set of mounted blades used to impart motion to a fluid (e.g., the rotor of a centrifugal pump).

impending blowout *n:* early manifestation or indication of a blowout.

impermeable *adj:* preventing the passage of fluid. A formation may be porous yet impermeable if there is an absence of connecting passages between the voids within it. See *permeability*.

impressed-current anode *n:* anode to which an external source of positive electricity is applied (as from a rectifier, DC generator, etc.) The negative electricity is applied to pipeline, casing, or other structure to be protected by the impressed-current method of cathodic protection.

Impeller

impression block *n:* a block with lead or another relatively soft material on its bottom. It is made up on drill pipe or tubing at the surface, run into a well, and allowed to rest on a tool or other object that has been lost in the well. When the block is retrieved, the size, shape, and position of the fish are obtained from the examination of the impression left in the lead, and an appropriate fishing tool may be selected.

in. *abbr:* inch.

in.^{2} *abbr:* square inch.

in.^{3} *abbr:* cubic inch.

incipient blowout *n:* See *impending blowout*.

independent *n:* a nonintegrated oil company whose operations are in the field of petroleum production, excluding transportation, refining, and marketing.

Independent Petroleum Association of America *n:* an organization of independent oil and gas producers headquartered in Tulsa, Oklahoma. Its function concerns the relationships between the oil industry and the public and government.

independent wire rope center *n:* a core for wire rope, consisting of a strand of steel wires with a spiral winding that is opposite that of the outer strands of the rope.

indicated volume *n:* the change in meter reading that occurs during a receipt or delivery of a liquid product.

indicated volume with factor *n:* the indicated volume multiplied by the meter factor for the particular liquid and operating conditions under which the meter was proved.

indirect heater *n:* apparatus or equipment in which heat from a primary source, usually the combustion of fuel, is transferred to a fluid or solid which acts as the heating medium.

induction survey *n:* an electric well log in which the conductivity of the formation rather than the resistivity is measured. Eddy currents are induced in the formations by a transmitter coil through which an alternating current circulates. The current sets up magnetic fields that induce voltages in a receiver coil. The voltages are amplified and recorded on the surface. Because oil-bearing formations are less conductive of electricity than water-bearing formations, an induction survey, when compared with resistivity readings, can aid in determination of oil and water zones. It is usually conducted in conjunction with a regular electric survey (ES), and thus its abbreviation is shortened to I-ES.

inelastic collision *n:* the collision of a neutron and the nucleus of an atom in which the total energy of the neutron is absorbed by the nucleus.

inert gas *n:* the part of a breathing medium that serves as a transport for oxygen and is not used by the body as a life-support agent. Its purpose is to dilute the flow of oxygen to the lungs, thereby preventing oxygen toxicity.

inertia *n:* the tendency of an object having mass to resist a change in velocity.

inertia brake *n:* a brake that utilizes the energy of a heavy, turning member to actuate the braking action.

infilling well *n:* a well drilled between known producing wells to better exploit the reservoir.

inhibited acid *n:* an acid that has been chemically treated before the acidizing or acid fracturing of a well to lessen its corrosive effect on the tubular goods and yet maintain its effectiveness. See *acid fracturing* and *acidize*.

inhibited mud *n:* a drilling fluid to which chemicals have been added to prevent it from causing clay particles in a formation to swell and thus impair the permeability of a productive zone. Salt is a mud inhibitor.

inhibitor *n:* an additive used to retard undesirable chemical action in a product; added in small quantity to gasolines to prevent oxidation and gum formation, to lubricating oils to stop color change, and to corrosive environments to decrease corrosive action.

initial potential *n:* the early production of an oilwell, recorded after testing operations and recovery of load oil and used as an indicator of the maximum ability of a well to produce on completion without subsequent reservoir damage.

initial set *n:* the point at which a cement slurry begins to harden, or set up, and is no longer pumpable.

initial stability *n:* the stability of an offshore drilling rig when upright or having only a small angle of heel.

injected gas *n:* a high-pressure gas injected into a formation to maintain or restore reservoir pressure; gas injected in gas-lift operations.

injection *n:* the process of forcing fluid into something. In a diesel engine, the introduction of high-pressure fuel oil into the cylinders.

injection log *n:* a survey used to determine the injection profile, that is, to assign specific volumes or percentages to each of the intervals taking fluid in

injection pattern-Instituto Mexicano del Petróleo

an injection well. The injection log is also used to check for casing or packer leaks, bad cement jobs, and fluid migration between zones.

injection pattern *n:* the spacing and pattern of wells in a secondary recovery or pressure-maintenance project, determined from the location of existing wells, type of offset operations used, reservoir size and shape, and cost of drilling new wells. Common injection patterns include line drive, five spot, seven spot, nine spot, and peripheral.

injection pump *n:* a chemical feed pump that injects chemical reagents into a flow-line system to treat emulsions, at a rate proportional to that of the flow of the well fluid. Operating power may come from electric motors or from linkage with the walking beam of a pumping well.

injection well *n:* a well in which fluids have been injected into an underground stratum to increase reservoir pressure.

inland barge rig *n:* a drilling structure consisting of a barge upon which the drilling equipment is constructed. When moved from one location to another, the barge floats, but, when stationed on the drill site, the barge is submerged to rest on the bottom. Typically, inland barge rigs are used to drill wells in marshes, shallow inland bays, and areas where the water covering the drill site is not too deep.

inlet manifold *n:* the passage that leads from the air filter to the cylinders of an engine. In a diesel engine, air only is introduced on the intake stroke.

innage *n:* the height of a liquid in a tank as measured from the bottom (datum plate) of the tank to the liquid surface.

innage gauge *n:* a measure of the liquid in a tank from the bottom of the tank to the surface of the liquid.

inorganic compounds *n pl:* chemical compounds that do not contain carbon as the principal element (excepting that in the form of carbonates, cyanides, and cyanates). Such compounds make up matter that is not plant or animal.

input shaft *n:* the transmission shaft for the drawworks that is driven directly by the compounding transmission on a mechanical-drive rig and is connected to it with the master clutch; or, on an electric-drive rig, the shaft driven directly by the electric motors. The input shaft drives the jackshaft or output shaft.

input well *n:* an injection well, used for injecting fluids into an underground stratum to increase reservoir pressure.

in./sec *abbr:* inches per second.

insert *n:* 1. a cylindrical object, rounded or chisel-shaped on one end and usually made of tungsten carbide, that is inserted in the cones of a bit, the cutters of a reamer, or the blades of a stabilizer to form the cutting element of the bit or the reamer or the wear surface of the stabilizer. Also called a compact. 2. a removable part molded to be set into the opening of the master bushing so that various sizes of slips may be accommodated. Also called a bowl.

insert pump *n:* a sucker rod pump that is run into the well as a complete unit. See *sucker rod pump.*

inside blowout preventer *n:* a valve installed in the drill stem to prevent a blowout through the stem. Flow is thus possible downward only, allowing mud to be pumped in but preventing any flow back up the stem. Also called an internal blowout preventer.

inside cutter *n:* See *internal cutter.*

inside diameter *n:* distance across the interior circle, especially in the measurement of pipe. See *diameter.*

in situ combustion *n:* a method of enhanced oil recovery in which heat is generated within the reservoir by injecting air and burning a portion of the oil in place. The heat of initial combustion cracks the crude hydrocarbons, vaporizes the lighter hydrocarbons, and deposits the heavier hydrocarbons as coke. As the fire moves from the injection well in the direction of producing wells, it burns the deposited coke, releases hot combustion gases, and converts connate water into steam. The vaporized hydrocarbons and the steam move ahead of the combustion zone, condensing into liquids as they cool and moving oil by miscible displacement and hot waterflooding. Combustion gases provide additional gas drive. Heat lowers the viscosity of the oil, causing it to flow more freely. This method is used to recover heavy, viscous oil. Also called fire flooding.

Instituto Mexicano del Petróleo (Mexican Petroleum Institute) *n:* a decentralized public-interest body created by the Mexican government. The main objective of the Instituto Mexicano del Petróleo is to carry out research and technological development as required by the petroleum, petrochemical, and chemical industries; to provide technical services for those industries; and to train personnel involved with the Mexican petroleum industry. The main offices of the Instituto Mexicano del Petróleo are located in México, D.F.

instrumentation *n:* a device or assembly of devices designed for one or more of the following functions: to measure operating variables (such as pressure, temperature, rate of flow, speed of rotation, etc.); to indicate these phenomena with visible or audible signals; to record them; to control them within a predetermined range; and to stop operations if the control fails. Simple instrumentation might consist of an indicating pressure gauge only. In a completely automatic system, desired ranges of pressure, temperature, and so on are predetermined and preset.

Instrument Society of America *n:* a group that sets standards for instruments made and used in the U.S.A.

insulating flange *n:* a flange equipped with plastic pieces to separate its metal parts, thus preventing the flow of electric current. Insulating flanges are often used in cathodic protection systems to prevent electrolytic corrosion and are sometimes installed when a flow line is being attached to a wellhead.

intake valve *n:* the cam-operated mechanism on an engine through which air and sometimes fuel are admitted to the cylinder.

integrating orifice meter *n:* an orifice meter with an automatic integrating device. It is constructed so that the product of the square roots of the differential and static pressures is recorded on the chart. The products are continuously totaled and shown on a counter index. When the product total is multiplied by the orifice flow constant, the rate of flow is determined directly.

interface *n:* the contact surface between two boundaries of liquids (e.g., the surface between water and the oil floating on it).

interfit *n:* the distance that the ends of one bit cone extend into the grooves of an adjacent one in a roller cone bit. Also called intermesh.

Intergovernmental Maritime Consultative Organization *n:* an international organization that regulates maritime practices, including possible pollution of the oceans by tanker-cleaning effluent.

intermediate casing string *n:* the string of casing set in a well after the surface casing is set to keep the hole from caving and to seal off troublesome formations. The string is sometimes called protection casing.

intermesh *n:* also called interfit. See *interfit.*

intermittent gas lift *n:* See *gas lift.*

intermitter *n:* a regulation device used in production of a flowing well. The well flows wide open (or through a choke) for short periods several times a day and is then closed in. Also used in some gas-lift installations.

internal blowout preventer *n:* also called inside blowout preventer. See *inside blowout preventer.*

internal-combustion engine *n:* a heat engine in which the pressure necessary to produce motion of the mechanism results from the ignition or burning of a fuel-air mixture within the engine cylinder.

internal cutter *n:* a fishing tool, containing metal-cutting knives, that is lowered into the length of pipe stuck in the hole to cut the pipe. The severed portion of the pipe can then be returned to the surface. Compare *external cutter.*

internal phase *n:* the fluid droplets or solids that are dispersed throughout another liquid in an emulsion. Compare *continuous phase.*

internal preventer *n:* also called an inside blowout preventer. See *inside blowout preventer.*

internal upset *n:* an extra-thick inside wall on the end of tubing or drill pipe at the point where it is threaded to compensate for the metal removed in threading. Unlike conventional drill pipe, which has the extra thickness on the outside, drill pipe with internal upset has the extra thickness inside and a uniform, straight wall outside. Compare *external upset.*

internal-upset pipe *n:* tubular goods in which the pipe walls at the threaded end are thickened (upset) on the inside to provide extra strength in the tool joints. Thus the outer wall of the pipe is the same diameter throughout its length. Upset casing is normally run at the top of long strings in deep operations.

International Association of Drilling Contractors *n:* a widely respected organization of drilling contractors headquartered in Houston, Texas. The organization sponsors or conducts research on education, accident prevention, drilling technology, and other matters of interest to drilling contractors and their employees. Formerly the American Association of Oilwell Drilling Contractors.

international system of units *n:* a system of units of measurement based on the metric system, adopted and described by the Eleventh General Conference on Weights and Measures. It provides an international standard of measurement to be followed when certain customary units, both U.S. and metric, are eventually phased out of international trade operations. The symbol *SI* (Le Système International d'Unités) designates the system, which involves seven base units: (1) metre for length, (2) kilogram for mass, (3) second for time, (4) kelvin for temperature, (5) ampere for electric current, (6) candela for luminous intensity, and (7) mole for amount of substance. From these units, others are derived without introducing numerical factors.

Interstate Commerce Commission n: a federal board that has jurisdiction over interstate pipelines.

interstice n: a pore space in a reservoir rock.

interstitial water n: water contained in the interstices of reservoir rock. In reservoir engineering, it is synonymous with connate water. See *connate water*.

intrusive rock n: an igneous rock that, while molten, penetrated into or between other rocks and solidified.

invaded zone n: an area within a permeable rock adjacent to a wellbore into which a filtrate (usually water) from the drilling mud has passed, with consequent partial or total displacement of the fluids originally present in the zone.

invert-emulsion mud n: an oil mud in which fresh or salt water is the dispersed phase and diesel, crude, or some other oil is the continuous phase. See *oil mud*.

ion n: an atom or a group of atoms charged either positively (a cation) or negatively (an anion) as a result of losing or gaining electrons.

ionization n: the process by which a neutral atom becomes positively or negatively charged through the loss or gain of electrons.

IPAA abbr: Independent Petroleum Association of America.

IR drop n: voltage drop, as determined by the formula $E = IR$.

iron count n: a measure of iron compounds in the product stream, determined by chemical analysis, that reflects the occurrence and the extent of corrosion.

iron roughneck n: manufacturer's term for a floor-mounted combination of a spinning wrench and a torque wrench. The iron roughneck moves into position hydraulically and eliminates the manual handling involved with suspended individual tools.

iron sponge process n: a method for removing small concentrations of hydrogen sulfide from natural gas by passing the gas over a bed of wood shavings which have been impregnated with a form of iron oxide. The impregnated wood shavings are called iron sponge. The hydrogen sulfide reacts with the iron oxide, forming iron sulfide and water.

ISA abbr: Instrument Society of America.

isobutane n: 1. a hydrocarbon of the paraffin series with the formula C_4H_{10} and having its carbon atoms branched. 2. in commercial transactions, a product meeting the GPA specification for commercial butane and, in addition, containing a minimum of 95 liquid volume percent isobutane.

isogonic chart n: a map that shows the isogonic lines joining points of magnetic declination, which is the variation between magnetic north and true north. For example, in Los Angeles, California, when the compass needle is pointing toward north, true north actually lies 15° east of magnetic north. See *declination*.

isogonic line n: an imaginary line on a map that joins places on the earth's surface at which the variation of a magnetic compass needle from true north is the same. This variation, which may range from 0 to 30 or more degrees either east or west of true north, must be compensated for to obtain an accurate reading of direction.

isomerization n: in petroleum refining, the process of altering the fundamental arrangement of the atoms in the molecule without adding or removing anything from the original material. Straight-chain hydrocarbons are converted to branched-chain hydrocarbons with a substantially higher octane rating in the presence of a catalyst at moderate temperatures and pressures. The process is basically important to the conversion of normal butene into isobutane.

isopach map n: a geological map of subsurface strata showing the various thicknesses of a given formation underlying an area. It is widely used in calculating reserves and in planning secondary recovery projects.

isopachous line n: one of the lines drawn on an isopach map to indicate areas of equal thickness in a stratigraphic unit.

isotope n: a form of an element that has the same atomic number as its other forms but has a different atomic mass. Isotopes of an element vary in the number of neutrons in the nucleus.

IWRC abbr: independent wire rope center.

J *sym:* joule.

jack *n:* 1. an oilwell pumping unit that is powered by an internal-combustion engine, electric motor, or rod line from a central power source. The walking beam of the pumping jack provides reciprocating motion to the pump rods of the well. See *walking beam.* 2. a device that is manually operated to turn an engine over for starting. *v:* to raise or lift.

jack board *n:* a device used to support the end of a length of pipe while another length is being screwed onto the pipe. It is sometimes referred to as a stabbing jack.

jacket *n:* tubular piece of steel in a tubing-liner type of sucker rod pump, inside of which is placed an accurately bored and honed liner. In this type of sucker rod pump, the pump plunger moves up and down within the liner, and the liner is inside the jacket.

jacket water *n:* water that fills, or is circulated through, a casing that partially or wholly surrounds a vessel or machine element in order to remove, add, or distribute heat and thereby to control the temperature within the vessel or element.

jackhammer *n:* 1. a rock drill that is pneumatically powered and usually held by the operator. 2. an air hammer.

jackknife mast *n:* a structural steel, open-sided tower raised vertically by special lifting tackle attached to the traveling block. See *mast.*

Jackup rig

jackknife rig *n:* a drilling rig that has a jackknife mast instead of a standard derrick.

jackshaft *n:* a short shaft that is usually set between two machines to provide increased or decreased flexibility and speed.

jackup *n:* a jackup drilling rig.

jackup drilling rig *n:* an offshore drilling structure with tubular or derrick legs that support the deck and hull. When positioned over the drilling site, the bottoms of the legs rest on the seafloor. A jackup rig is towed or propelled to a location with its legs up. Once the legs are firmly positioned on the bottom, the deck and hull height are adjusted and leveled.

jar *n:* a percussion tool operated mechanically or hydraulically to deliver a heavy hammer blow to objects in the borehole. Jars are used to free objects stuck in the hole or to loosen tubing or drill pipe that is hung up. Blows may be delivered downward or upward, the jar being controlled at the surface. *v:* to apply a heavy blow to the drill stem by use of a jar.

jar accelerator *n:* a hydraulic tool used in conjunction with a jar and made up on the fishing string above the jar to increase the power of the hammer blow.

jaw clutch *n:* a positive-type clutch in which one or more jaws mesh together in the opposing clutch sections.

jerk line *n:* a short rope used on the cathead of the drilling rig to tighten pipe joints by pulling on the makeup tongs. See *makeup tongs.*

jet *n:* 1. a hydraulic device operated by pump pressure to clean mud pits and tanks in rotary drilling and to mix mud components. 2. in a perforating gun using shaped charges, a highly penetrating, fast-moving stream of exploded particles that cuts a hole in the casing, cement, and formation.

jet bit *n:* a drilling bit having replaceable nozzles through which the drilling fluid is directed in a high-velocity stream to the bottom of the hole to improve the efficiency of the bit. See *bit.*

jet compressor *n:* a device employing a venturi nozzle so that a high-pressure stream flowing through the nozzle creates a lower pressure or a vacuum into which the gas to be compressed flows. The gas is discharged from the nozzle with the expanded high-pressure medium.

jet cutoff *n:* a procedure for severing pipe stuck in a well by detonating special shaped-charge explosives similar to those used in jet perforating. The explosive is lowered into the pipe to the desired depth and detonated. The force of the explosion makes radiating horizontal cuts around the pipe, and the severed portion of the pipe is retrieved.

jet deflection bit *n:* a special jet bit that has a very large nozzle used to deflect a hole from the vertical. The large nozzle erodes away one side of the hole so that the hole is deflected off vertical. A jet deflection bit is especially effective in soft formations.

jet gun *n:* an assembly, including a carrier and shaped charges, that is used in jet perforating.

jet out *v:* to use a jet for cleaning out mud tanks, cellar, and other areas.

jet-perforate *v:* to create a hole through the casing with a shaped charge of high explosives instead of a gun that fires projectiles. The loaded charges are lowered into the hole to the desired depth. Once detonated, the charges emit short, penetrating jets of high-velocity gases that cut holes in the casing and cement and some distance into the formation. Formation fluids then flow into the wellbore through these perforations. See *bullet perforator* and *gun-perforate.*

jet pump *n:* a pump that operates by means of a jet of steam, water, or other fluid that imparts motion and subsequent pressure to a fluid medium.

jobber *n:* a wholesaler who buys gasoline for resale to retailers.

joint *n:* a single length (30 feet or 9 m) of drill pipe, drill collar, casing, or tubing that has threaded connections at both ends. Several joints screwed together constitute a stand of pipe.

joint identifier *n:* a gauge for determining whether the connections of drill collars and tool joints match.

joint movement *n:* the shipment of a tender of oil through the facilities of one or more pipeline companies.

joint tariff *n:* a rate sheet, issued jointly by two or more companies, setting forth charges for moving oil over the facilities of each.

joule *n:* the unit used to measure heat, work, and energy in the metric system. Its symbol is J. It is the amount of energy required to move an object of 1 kilogram mass to a height of 1 metre. Also called a newton-metre.

Joule-Thomson effect *n:* the change in gas temperature which occurs when the gas is expanded adiabatically from a higher pressure to a lower pressure. The effect for most gases, except hydrogen and helium, is a cooling of the gas.

journal *n:* the part of a rotating shaft that turns in a bearing.

journal angle *n:* the angle formed by lines perpendicular to the axis of the journal and the axis of the bit. Also called pin angle.

journal bearing *n:* a machine part in which a rotating shaft (a journal) revolves or slides.

J-tool *n:* a sleeve receptacle that has a fitted male element and pins that fit into milled J-shaped slots on the sleeve. The short sides of the J-slots provide a shoulder for supporting weight on the pins of the male element. When the male element is lowered and turned relative to the sleeve, the pins slide in the slot towards the long side of the J, which is open-ended. The pins may thus be raised out, releasing weight that may be supported by the sleeve. The releasing procedure is called "unjaying the tool."

jug *n:* See *geophone*.

jug hustler *n:* (slang) the member of a seismograph crew who places the geophones.

jumbo burner *n:* a flare for burning waste gas when the volume of gas is very small or when no market is readily available.

jumbo tank cars *n pl:* tank cars having capacities of 30,000 gallons (114 m³) or more. Standard tank cars have a capacity of 10,000 to 11,000 gallons (38-42m³).

junior orifice fitting *n:* a one-piece orifice fitting without flanges.

junk *n:* metal debris lost in a hole. Junk may be a lost bit, pieces of a bit, milled pieces of pipe, wrenches, or any relatively small object that impedes drilling and must be fished out of the hole. *v:* to abandon (as a nonproductive well).

junk basket *n:* a device made up on the bottom of the drill stem to catch pieces of junk from the bottom of the hole. Mud circulation forces the junk into a barrel in the tool, where it is held by metal projections, or catchers. When the basket is brought back to the surface, the junk is removed. Also called a junk sub.

junk sub *n:* also called a junk basket. See *junk basket*.

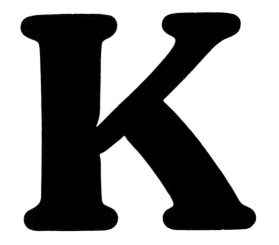

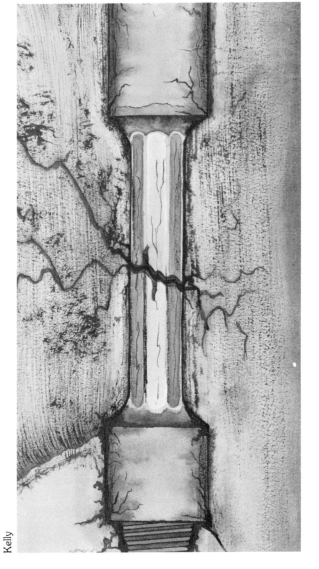

Kelly

K *sym:* kelvin.

KB *abbr:* kelly bushing; used in drilling reports.

K capture *n:* an interaction in which a nucleus captures an electron from the K shell of atomic electrons (the shell nearest the nucleus) and emits a neutrino.

keel *n:* a centerline strength member running fore and aft along the bottom of a ship or a floating offshore drilling rig, forming the backbone of the structure.

kelly *n:* the heavy steel member, three-, four-, six-, or eight-sided, suspended from the swivel through the rotary table and connected to the topmost joint of drill pipe to turn the drill stem as the rotary table turns. It has a bored passageway that permits fluid to be circulated into the drill stem and up the annulus, or vice versa.

kelly bushing *n:* a special device that, when fitted into the master bushing, transmits torque to the kelly and simultaneously permits vertical movement of the kelly to make hole. It may be shaped to fit the rotary opening or have pins for transmitting torque. Also called the drive bushing. See *kelly*.

kelly bushing rollers *n:* rollers in the kelly drive bushing that roll against the flat sides of the kelly and allow the kelly to move freely upward or downward. Also called drive rollers.

kelly cock *n:* a valve installed at one or both ends of the kelly. When a high-pressure backflow begins inside the drill stem, the valve is closed to keep pressure off the swivel and rotary hose.

kelly flat *n:* one of the flat sides of a kelly. Also called a flat. See *kelly*.

kelly hose *n:* also called the mud hose or rotary hose. See *rotary hose*.

kelly joint *n:* See *kelly*.

kelly saver sub *n:* a sub that fits in the drill stem between the kelly and the drill pipe. Threads on the drill pipe mate with those of the sub, minimizing wear on the kelly.

kelly spinner *n:* a pneumatically operated device mounted on top of the kelly that, when actuated, causes the kelly to turn or spin. It is useful when the kelly or a joint of pipe attached to it must be spun, that is, rotated rapidly for being made up.

kelvin *n:* the fundamental unit of thermodynamic temperature in the metric system. The symbol for kelvin is K. See *Kelvin temperature scale*.

Kelvin temperature scale *n:* a temperature scale with the degree interval of the Celsius scale and the zero point at absolute zero. On the Kelvin scale, water freezes at 273 K and boils at 373 K. See *absolute temperature scale*.

kerosine *n:* a light, flammable hydrocarbon fuel or solvent. It is also spelled *kerosene*, but *kerosine* is preferred by the API to parallel *gasoline*.

key *n:* 1. a hook-shaped wrench that fits the square shoulder of a sucker rod and is used when rods are pulled or run into a pumping oilwell. Usually used in pairs; one key backs up and the other breaks out or makes up the rod. Also called a rod wrench. 2. a slender strip of metal that is used to fasten a wheel or a gear onto a shaft. The key fits into slots in the shaft and in the wheel or gear.

key *v:* to use a cotter key to prevent a nut from coming loose from a bolt or a stud.

key seat *n:* 1. a channel or groove cut in the side of the hole of a well and parallel to the axis of the hole. A key seat results from the dragging of pipe on a sharp bend in the hole. 2. a groove cut parallel to the axis in a shaft or a pulley bore.

key seat barge *n:* a barge in which the mast is placed over a channel cut out of the side of the barge and through which drilling or workover operations are performed.

key seat wiper *n:* a device used to ream out a hole where key seating has occurred. Usually a jar is used first to loosen stuck pipe from a sharp bend, and then a key seat wiper is used to enlarge the hole at the key seat caused by the pipe. See *key seat*.

kg *abbr:* kilogram.

kick *n:* an entry of water, gas, oil, or other formation fluid into the wellbore during drilling. It occurs because the pressure exerted by the column of drilling fluid is not great enough to overcome the pressure exerted by the fluids in the formation drilled. If prompt action is not taken to control the kick or kill the well, a blowout may occur.

kick off *v:* 1. to bring a well into production; used most often when gas is injected into a gas-lift well to start production. 2. in workover operations, to swab a well to restore it to production. 3. to deviate a wellbore from the vertical, as in directional drilling.

kickoff point *n:* the depth in a vertical hole at which a deviated or slant hole is started; used in directional drilling.

kickoff pressure *n:* the gas pressure required to kick off a gas-lift well, usually greater than that required to maintain the well in production. See *gas lift*.

kill *v:* 1. in drilling, to prevent a threatened blowout by taking suitable preventive measures (e.g., to shut in the well with the blowout preventers, circulate the kick out, and increase the weight of the drilling mud). 2. in production, to stop a well from producing oil and gas so that reconditioning of the well can proceed. Production is stopped by circulating water and mud into the hole.

kill line *n:* a high-pressure line that connects the mud pump and the well and through which heavy drilling fluid can be pumped into the well to control a threatened blowout.

kill sheet *n:* a printed form that contains blank spaces for recording information about killing an impending blowout, provided to remind personnel of the necessary steps to kill a well.

kilogram *n:* the metric unit of mass equal to 1 000 grams. Its symbol is kg.

kilopascal *n:* 1 000 pascals. See *pascal*.

kilowatt *n:* a metric unit of power equal to approximately 1.34 horsepower.

kinematic viscosity *n:* the absolute viscosity of a fluid divided by the density of the fluid at the temperature of the viscosity measurement.

kink *n:* a loop in a wire rope that, having been pulled tight, causes permanent distortion of the wire rope.

Kirchoff's second law *n:* the law stating that, at each instant of time, the increases in voltage around a closed loop in a network is equal to the algebraic sum of the voltage drops.

knockout *n:* any liquid condensed from a stream by a scrubber following compression and cooling.

knockout drops *n:* (slang) a slugging compound.

knot *n:* a unit of speed equal to 1 nautical mile (6,020.2 feet) per hour. It is also equal to about $1\frac{1}{7}$ statute miles per hour.

knowledge box *n:* (slang) the cupboard or desk in which the driller keeps the various records pertaining to a drilling operation.

knuckle joint *n:* a deflection tool, placed above the drill bit in the drill stem, with a ball and socket arrangement that allows the tool to be deflected at an angle; used in directional drilling. A knuckle joint is useful in fishing operations because it allows the fishing tool to be deflected to the side of the hole where a fish may have come to rest.

KO *abbr:* kicked off; used in drilling reports.

KOP *abbr:* kickoff point.

kPa *sym:* kilopascal.

K shell *n:* the shell of electrons nearest the nucleus in an atom.

kV *sym:* kilovolt.

K value *n:* See *vapor-liquid equilibrium ratio.*

kW *sym:* kilowatt.

kwh, KWH, kw-h *abbr:* kilowatt-hour.

LACT *abbr:* lease automatic custody transfer.

LACT unit *n:* an automated system for measuring, testing, and transferring oil from a lease gathering system into a pipeline. See *lease automatic custody transfer.*

laminar flow *n:* a smooth flow of fluid in which no cross flow of fluid particles occurs between adjacent stream lines.

land *n:* 1. the area of a partly machined surface (as with grooves or indentation) that is left smooth. 2. the area on a piston between the grooves into which the rings fit.

land casing *v:* to install casing so that it is supported in the casinghead by slips. The casing is usually landed in the casinghead at exactly the position in which it was hanging when the cement plug reached its lowest point.

landing depth *n:* the depth to which the lower end of casing extends in the hole when casing is landed.

landman *n:* a person in the petroleum industry who negotiates with landowners for land options, oil drilling leases, and royalties and with producers for the pooling of production in a field; also called a leaseman.

land rig *n:* any drilling rig that is located on dry land. Compare *offshore rig.*

lanolin *n:* wool grease, derived from the preparation of raw wool for spinning; it is used in cosmetics, shampoos, and a variety of industrial products.

lap *n:* an interval in the cased hole where the top of a liner overlaps the bottom of a string of casing.

last engaged thread *n:* the last pipe thread that is actually screwed into the coupling thread in making up a joint of drill pipe, drill collars, tubing, or casing. If the pipe makes up perfectly, it is also the last thread cut on the pipe.

latch on *v:* to attach elevators to a section of pipe to pull it out of or run it into the hole.

latex cement *n:* an oilwell cement composed of latex, cement, a surfactant, and water and characterized by its high-strength bond with other materials and its resistance to contamination by oil or drilling mud.

lay *n:* 1. the spiral of strands in a wire rope either to the right or to the left, as viewed from above. 2. a term used to measure wire rope, signifying the linear distance a wire strand covers in one complete rotation around the rope.

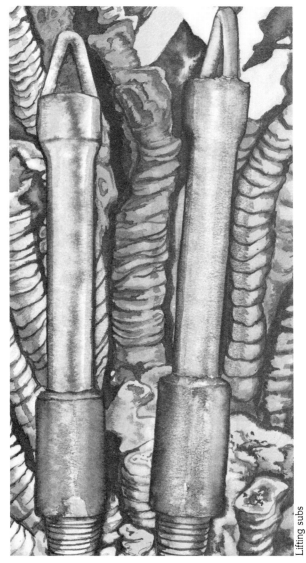

Lifting subs

lay barge *n:* a barge used in the construction and placement of underwater pipelines. Joints of pipe are welded together and then lowered off the stern of the barge as it moves ahead.

lay down pipe *v:* to pull drill pipe or tubing from the hole and place it in a horizontal position on a pipe rack. Compare *set back.*

layer *n:* a bed, or stratum, of rock.

laying down *n:* the operation of laying down pipe. See *lay down pipe.*

lb *abbr:* pound.

lb/ft³ *abbr:* pounds per cubic foot.

lead acetate test *n:* a method for detecting the presence of hydrogen sulfide in a fluid by discoloration of paper which has been moistened with lead acetate solution.

lead line *n:* the pipe through which oil or gas flows from a well to additional equipment on the lease.

lead-tong man *n:* the crew member who operates lead tongs during hoisting of the drill pipe.

lead tongs *n pl:* (pronounced "leed") the pipe tongs suspended in the derrick or mast and operated by a wireline connected to the breakout cathead. In coming out of the hole, they are used on the pin end of the joint for breaking out. In going into the hole, they are used on the box end as backup to the makeup tongs.

leak-off rate *n:* the rate at which a fracturing fluid leaves the fracture and enters the formation surrounding the fracture. Generally, it is desirable for fracturing fluids to have a low leak-off rate (i.e., very little fluid should enter the formation being fractured) so that the fracture can be better extended into the formation.

leak-off test *n:* a gradual pressurizing of the casing after the blowout preventers have been installed to permit estimation of the formation fracture pressure at the casing seat.

lean amine *n:* amine solution that has been stripped of absorbed acid gases, giving a solution suitable for recirculation to the contactor.

lean gas *n:* 1. residue gas remaining after recovery of natural gas liquids in a gas processing plant. Also called dry gas. 2. unprocessed gas containing few or no recoverable natural gas liquids.

lean oil *n:* a hydrocarbon liquid, usually lighter in weight than kerosine and heavier than paint thinner. In a gas processing plant, lean oil is used in an absorber to remove heavier hydrocarbons from natural gas.

lease *n:* 1. a legal document executed between a landowner, as lessor, and a company or individual, as lessee, that grants the right to exploit the premises for minerals or other products. 2. the area where production wells, stock tanks, separators, LACT units, and other production equipment are located.

lease automatic custody transfer *n:* the measurement, testing, and transfer of oil from the producer's tanks to the connected pipeline on an automatic basis without a representative of either the producer or the gathering company having to be present. See *LACT unit.*

lease hound *n:* (slang) a landman who procures leases on tracts of land for petroleum exploration and production.

leaseman *n:* See *landman.*

lease metering site *n:* the point on a lease where the volume of oil produced from the lease is measured, usually automatically.

lease operator *n:* also called a pumper. See *pumper.*

lease tank *n:* also called a production tank. See *production tank.*

leeward *adj:* (nautical) downwind.

lens *n:* 1. a porous, permeable, irregularly shaped sedimentary deposit surrounded by impervious rock. 2. a lenticular sedimentary bed that pinches out, or comes to an end, in all directions.

lens-type trap *n:* a hydrocarbon reservoir consisting of a porous, permeable, irregularly shaped sedimentary deposit surrounded by impervious rock. See *lens.*

lessee *n:* the recipient of a lease (such as an oil and gas lease).

lessor *n:* the conveyor of a lease (such as an oil and gas lease).

level *n:* 1. the height or depth at which the top of a column of fluid is located (as the level of fluid in a well). 2. a horizontally even surface. 3. a device used to determine whether a surface is horizontal.

lifeline *n:* a line attached to a diver's helmet by which he is lowered and raised in the water.

lifter-roof tank *n:* a tank whose roof rises and falls with the changes of pressure in the tank but does not float on the product stored in it.

lifting nipple *n:* a short piece of pipe with a pronounced upset, or shoulder, on the upper end, screwed into drill pipe, drill collars, or casing to provide a positive grip for the elevators; also called a lifting sub or a hoisting plug.

lifting sub *n:* also called hoisting plug or lifting nipple. See *lifting nipple*.

light crude oil *n:* a crude oil of relatively high API gravity (usually 40 degrees or higher).

light displacement *n:* on mobile offshore drilling rigs, the weight of the rig with all permanently attached equipment but without fuel, supplies, crew, ballast, drill pipe, and so forth.

light ends *n pl:* the lighter hydrocarbon molecules that comprise gasoline, light kerosine, heptane, natural gas, and so forth.

lightening hole *n:* a hole cut into a strengthening member that reduces its weight but does not significantly affect its strength.

light hydrocarbons *n pl:* the low molecular weight hydrocarbons such as methane, ethane, propane, and butanes.

lightweight cement *n:* a cement or cement system that handles stable slurries having a density less than that of neat cement. Lightweight cements are used in low-pressure zones where the high hydrostatic pressure of long columns of neat cement can fracture the formation and result in lost circulation.

lightweight gear *n:* all diving equipment less complex than the standard dress. This equipment employs face masks or helmets, protective clothing, and swim fins or boots.

lignosulfonate *n:* an organic drilling fluid additive derived from by-products of a paper-making process using sulfite; added to drilling mud to minimize fluid loss and to reduce viscosity of the mud.

limber hole *n:* a hole cut in a structural member of a ship or offshore drilling rig, usually in a tank, to allow water to pass through freely.

lime *n:* a caustic solid that consists primarily of calcium oxide (CaO). Many forms of CaO are called lime, including the various chemical and physical forms of quicklime, hydrated lime, and even calcium carbonate. Limestone is sometimes called lime.

lime mud *n:* a drilling mud that is treated with lime to provide a source of soluble calcium in the filtrate in order to obtain desirable mud properties for drilling in shale or clay formations.

limestone *n:* a sedimentary rock rich in calcium carbonate that sometimes serves as a reservoir rock for petroleum.

limited-entry technique *n:* a fracturing method in which fracturing fluid is injected into the formation through a limited number of perforations (i.e., fluid is not injected through all the perforations at once; rather, injection is confined to a few selected perforations). This special technique can be useful when long, thick, or multiple producing zones are to be fractured.

line *n:* 1. any length of pipe through which liquid or gas flows. 2. rope or wire rope. 3. electrical wire.

linear polarization *n:* a technique used to measure instantaneous corrosion rates by changing the electrical potential of a structure that is corroding in a conductive fluid and measuring the current required for that change.

line drive *n:* in waterflooding, a straight-line pattern of injection wells designed to advance water to the producing wells in the form of a nearly linear frontal movement. See *waterflood*.

line pipe *n:* a steel or plastic pipe used in pipelines, gathering systems, flow lines, and so forth.

liner *n:* 1. a string of casing used to case open hole below existing casing. Liner casing extends from the setting depth up into another string of casing, usually overlapping about 100 feet above the lower end of the intermediate or oil string. Liners are nearly always suspended from the upper string by a hanger device. 2. in jet perforating guns, a conically shaped metallic piece that is part of a shaped charge. It increases the efficiency of the charge by increasing the penetrating ability of the jet. 3. a replaceable tube that fits inside the cylinder of an engine or a pump.

liner barrel *n:* a pump barrel used for either tubing pumps or rod (insert) pumps. A full-cylinder barrel consists of a steel jacket inside of which is a full-length tube of cast iron or special alloy. The inner surface of the barrel is polished to a mirrorlike finish to permit a fluid-tight seal between it and the plunger. In a sectional liner barrel, the tube placed inside the steel jacket consists of a series of sections placed end to end and held firmly in place by means of threaded collars on the ends of the steel jacket.

liner completion *n:* a well completion in which a liner is used to obtain communication between the reservoir and the wellbore.

liner hanger n: a slip device that attaches the liner to the casing. See *liner*.

liner lap n: the distance that a liner extends into the bottom of a string of casing.

liner patch n: a stressed-steel corrugated tube that is lowered into existing casing in a well in order to repair a hole or leak in the casing. The patch is cemented to the casing with glass fiber and epoxy resin.

line scraper n: also called a pig. See *pig*.

line spooler n: a rubberlike device fitted on the drawworks and used to cause the fastline to reverse its direction on the drawworks drum or spool when a layer of line is completed on the drum and the next layer is started.

link ear n: a steel projection on the drilling hook by means of which the elevator links are attached to the hook.

lipophilic adj: having an affinity for lipids, a class of compounds that includes most hydrocarbons.

liquefaction n: the process whereby a substance in its gaseous or solid state is liquefied.

liquefied natural gas n: a liquid composed chiefly of natural gas (i.e., mostly methane). Natural gas is liquefied to make it easy to transport if a pipeline is not feasible (as across a body of water). Not as easily liquefied as LPG, LNG must be put under low temperature and high pressure or under extremely low (cryogenic) temperature and close to atmospheric pressure to become liquefied.

liquefied petroleum gas n: a mixture of heavier, gaseous, paraffinic hydrocarbons, principally butane and propane. These gases, easily liquefied at moderate pressure, may be transported as liquids but converted to gases on release of the pressure. Thus, liquefied petroleum gas is a portable source of thermal energy that finds wide application in areas where it is impractical to distribute natural gas. It is also used as a fuel for internal-combustion engines and has many industrial and domestic uses. Principal sources are natural and refinery gas, from which the liquefied petroleum gases are separated by fractionation.

liquefied refinery gas n: liquid propane or butane produced by a crude oil refinery. It may differ from LP gas in that propylene and butylene may be present.

liquid n: a state of matter in which the shape of the given mass depends on the containing vessel, but the volume of the mass is independent of the vessel. A liquid is a fluid that is almost incompressible.

liquid and solid ROB (remaining on board) n: the measurable material remaining on board a vessel after a discharge. Includes measurable sludge, sediment, oil, and water or oily residue lying on the bottom of the vessel's cargo compartments and in associated lines and pumps.

liquid desiccant n: a hygroscopic liquid, such as glycol, used to remove water from other fluids.

liquid-level controller n: any device used to control the liquid level in a tank by actuating electric or pneumatic switches that open and close the discharge valve or the intake valve, thus maintaining the liquid at the desired level.

liquid-level gauge n: any device that indicates the level or quantity of liquid in a container.

liquid-level indicator n: a device connected to a vessel, coupled with either a float in the vessel or directly with the fluid therein, and calibrated to give a visual indication of the liquid level.

list n: the position of a ship or offshore drilling rig that heels to one side because of a shift in cargo, machinery, or supplies.

lithification n: the conversion of unconsolidated deposits into solid rock.

lithology n: 1. the study of rocks, usually macroscopic. 2. the individual character of a rock in terms of mineral composition, structure, and so forth.

litre n: a unit of metric measure of capacity equal to the volume occupied by 1 kg of water at 4°C and at the standard atmospheric pressure of 760 mm.

liveboating n: a diving operation involving the use of a boat or vessel that is underway.

lm abbr: lime; used in drilling reports.

LNG abbr: liquefied natural gas.

LNGC abbr: liquefied natural gas carrier.

load n: 1. in mechanics, the weight or pressure placed on an object. The load on a bit refers to the amount of weight of the drill collars allowed to rest on the bit. See *weight on the bit*. 2. in reference to engines, the amount of work that an engine is doing—for example, 50 percent load means that the engine is putting out 50 percent of the power that it is able to produce. v: to engage an engine so that it works. Compare *idle*.

load binder n: a chain or cable with a latching device, used to secure loads (usually of pipe) on trucks. It is also called a boomer.

loader *n:* the individual who handles the filling of tank cars or transport trucks.

load guy *n:* See *guy line.*

load guy line *n:* the wire rope attached to a mast or derrick to provide the main support for the structure. Compare *wind guy line.*

loading rack *n:* the equipment used for transferring crude oil or petroleum products into tank cars or trucks.

load oil *n:* the crude or refined oil used in fracturing a formation to stimulate a well, as distinguished from the oil normally produced by the well.

LOC *abbr:* location; used in drilling reports.

location *n:* the place where a well is drilled; also called well site.

log *n:* a systematic recording of data, such as a driller's log, mud log, electrical well log, or radioactivity log. Many different logs are run in wells to obtain various characteristics of downhole formations. *v:* to record data.

log a well *v:* to run any of the various logs used to ascertain downhole information about a well.

logbook *n:* a book used by station engineers, dispatchers, and gaugers for keeping notes on current operating data.

logging devices *n:* any of several electrical, acoustical, mechanical, or radioactivity devices that are used to measure and record certain characteristics or events that occur in a well that has been or is being drilled.

longitude *n:* the arc or portion of the earth's equator intersected between the meridian of a given place and the prime meridian (at Greenwich, England) and expressed either in degrees or in time.

long string *n:* 1. the last string of casing set in a well. 2. the string of casing that is set at the top of or through the producing zone, often called the oil string or production casing.

loss of circulation *n:* See *lost circulation.*

lost circulation *n:* the quantities of whole mud lost to a formation, usually in cavernous, fissured, or coarsely permeable beds, evidenced by the complete or partial failure of the mud to return to the surface as it is being circulated in the hole. Lost circulation can lead to a blowout and, in general, reduce the efficiency of the drilling operation. Also called lost returns.

lost circulation material *n:* a substance added to cement slurries or drilling mud to prevent the loss of cement or mud to the formation. See *bridging material.*

lost circulation plug *n:* cement set across a formation that is taking excessively large amounts of drilling fluid during drilling operations.

lost hole *n:* a well that cannot be further drilled or produced because of a blowout, unsuccessful fishing job, and so forth.

lost pipe *n:* drill pipe, drill collars, tubing, or casing that has become separated in the hole from the part of the pipe reaching the surface, necessitating its removal before normal operations can proceed; a fish.

lost returns *n:* also called lost circulation. See *lost circulation.*

LOT (loaded on top) procedure *n:* a procedure in which tank-cleaning operations are carried out on board ships and the resulting water/oil mixture is collected in a tank and allowed to separate. The relatively clean water is then pumped out of the vessel, and part of the next cargo is loaded on top of the remaining cargo/water-cleaning residue. This residue is called slops.

low drum drive *n:* the drive for the drawworks drum used when hoisting loads are heavy.

lower kelly cock *n:* also called drill stem safety valve. See *drill stem safety valve.*

lower tier *n:* a category of oil production for purposes of price control. Lower tier refers to *old oil*, oil contained in reservoirs that were being produced during 1972.

low-solids mud *n:* a drilling mud that contains a minimum amount of solid material (sand, silt, etc.) and is used in rotary drilling when possible because it can provide fast drilling rates.

low-temperature fractionation *n:* separation of a hydrocarbon fluid mixture into components by fractionation, wherein the reflux condenser is operated at temperatures requiring refrigeration. See *Pod analysis.*

low-temperature processing *n:* gas processing conducted below ambient temperatures.

LPG *abbr:* liquefied petroleum gas.

LRG *abbr:* liquefied refinery gas.

ls *abbr:* limestone; used in drilling reports.

lse *abbr:* lease; used in drilling reports.

lubricant *n:* a substance – usually petroleum-based – that is used to reduce friction between two moving parts.

lubricator n: a specially fabricated length of casing or tubing usually placed temporarily above a valve on top of the casinghead or tubing head; used to run swabbing or perforating tools into a producing well; provides method for sealing off pressure and thus should be rated for highest anticipated pressure.

lug n: a projection on a casting to which a bolt or other part may be fitted.

lugging power n: the torque, or turning power, delivered to the flywheel of a diesel engine.

m *sym:* metre.

m² *sym:* square metre.

m³ *sym:* cubic metre.

mA *sym:* milliampere.

macaroni rig *n:* a workover rig, usually lightweight, that is specially built to run a string of ¾-inch or 1-inch tubing. See *macaroni string*.

macaroni string *n:* a string of tubing or pipe, usually ¾ or 1 inch in diameter.

magma *n:* the hot fluid matter within the earth's crust that is capable of intrusion or extrusion and that produces igneous rock when cooled.

Magnaflux *n:* trade name for the equipment and processes used for detecting cracks and other surface discontinuities in iron or steel. A magnetic field is set up in the part to be inspected, and a powder or paste of magnetic particles is applied. The particles arrange themselves around discontinuities in the metal, revealing defects.

magnetic brake *n:* also called an electrodynamic brake. See *electrodynamic brake*.

magnetic surveying instrument *n:* a device used to determine the direction and drift of a deviated wellbore. It uses a plumb bob, a magnetic compass, and photographic or mechanical equipment to determine and record directional information. See *directional survey* and *directional drilling*.

magnetostrictive transducer *n:* a tightly banded scroll of special steel which vibrates when a

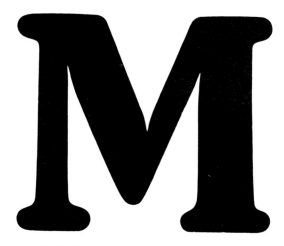

magnetic field is applied to it. The vibration of the scroll sets off a transmitter which is alternately switched on and off at 15 to 60 times per second. The transmitter causes compressional sound waves to travel through the formations surrounding the borehole. Each different formation material will exhibit its own characteristic effect upon the elastic wave propagation.

main deck *n:* the principal deck extending from front to back of a ship or offshore drilling rig; also called the Texas deck.

mainline *n:* a large-diameter pipeline between distant points; a trunk line.

mainline plant *n:* a plant that processes the gas which is being transported through a cross-country

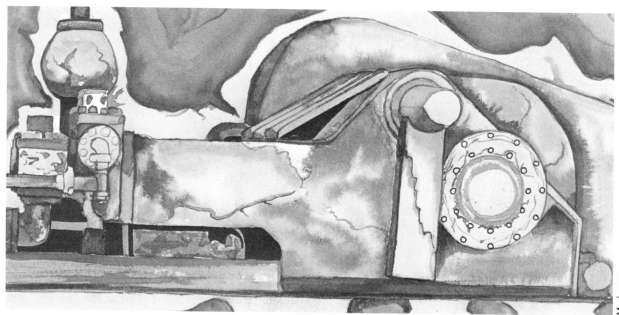

Mud pump

transmission line. Also called pipeline, on-line, or straddle plant.

Majors *n pl:* seven oil companies that, until recently, controlled a huge percentage of the development, production, refining, transport, and marketing of the international oil industry. The seven Majors are Exxon, Chevron, Mobil Oil, Gulf Oil, Texaco, Royal Dutch/Shell, and British Petroleum. Nowadays, the Majors continue to control about half of the world's development, refining, and marketing.

make a connection *v:* to attach a joint of drill pipe onto the drill stem suspended in the wellbore to permit deepening the wellbore by the length of the joint added (30 feet or 9 m).

make a hand *v:* (slang) to become a good worker.

make a trip *v:* to hoist the drill stem out of the wellbore to perform one of a number of operations such as changing bits, taking a core, and so forth, and then to return the drill stem to the wellbore.

make hole *v:* to deepen the hole made by the bit; to drill ahead.

makeup *adj:* added to a system (as makeup water used in mixing mud).

make up *v:* 1. to assemble and join parts to form a complete unit (as to make up a string of casing). 2. to screw together two threaded pieces. 3. to mix or prepare (as to make up a tank of mud). 4. to compensate for (as to make up for lost time.)

make up a joint *v:* to screw a length of pipe into another length of pipe.

makeup cathead *n:* a device that is attached to the shaft of the drawworks and used as a power source for screwing together joints of pipe; usually located on the driller's side of the drawworks. Also called spinning cathead. See *cathead*.

makeup gas *n:* 1. gas that is taken in succeeding years and has been paid for previously under a take-or-pay clause in a gas purchase contract. The contract will normally specify the number of years after payment in which the purchaser can take delivery or makeup gas without paying a second time. 2. gas injected into a reservoir to maintain a constant reservoir pressure and thereby prevent retrograde condensation. 3. in gas processing, the gas that makes up for plant losses. During processing there is a reduction in gas volume because of fuel and shrinkage. Some agreements between gas transmission companies and plant owners require plant losses to be made up or to be paid for.

makeup tongs *n:* tongs used for screwing one length of pipe into another for making up a joint. See *tongs* and *lead tongs*.

male connection *n:* a pipe, coupling, or tool that has threads on the outside so that it can be joined to a female connection.

mandrel *n:* a cylindrical bar, spindle, or shaft, around which other parts are arranged or attached or which fits inside a cylinder or tube.

manhole *n:* a hole in the top or side of a tank through which a person can enter.

manifold *n:* 1. an accessory system of piping to a main piping system (or another conductor) that serves to divide a flow into several parts, to combine several flows into one, or to reroute a flow to any one of several possible destinations. 2. a pipe fitting with several side outlets to connect it with other pipes. 3. a fitting on an internal-combustion engine made to receive exhaust gases from several cylinders.

manometer *n:* a U-shaped piece of glass tubing containing a liquid (usually water or mercury) that is used to measure the pressure of gases or liquids. When pressure is applied, the liquid level in one arm rises while the level in the other drops. A set of calibrated markings beside one of the arms permits a pressure reading to be taken, usually in inches or millimetres.

marginal well *n:* a well that is approaching depletion of its natural resource to the extent that any profit from continued production is doubtful.

marine riser connector *n:* a fitting on top of the subsea blowout preventers to which the riser pipe is connected.

marine riser system *n:* See *riser pipe*.

marker bed *n:* a distinctive, easily identified rock stratum, especially one used as a guide for drilling or correlation of logs.

marl *n:* a semisolid or unconsolidated clay, silt, or sand.

Marsh funnel *n:* a calibrated funnel used in field tests to determine the viscosity of drilling mud.

mast *n:* a portable derrick that is capable of being erected as a unit, as distinguished from a standard derrick that cannot be raised to a working position as a unit. For transporting by land, the mast can be divided into two or more sections to avoid excessive length extending from truck beds on the highway. Compare *derrick*.

master bushing n: a device that fits into the rotary table. It accommodates the slips and drives the kelly bushing so that the rotating motion of the rotary table can be transmitted to the kelly. Also called rotary bushing.

master clutch n: the clutch that connects the compounding transmission on a mechanical-drive rig to the input shaft to the drawworks.

master gate n: also called master valve. See *master valve*.

master valve n: 1. a large valve located on the Christmas tree and used to control the flow of oil and gas from a well. Also called a master gate. 2. the blind or blank rams of a blowout preventer.

matrix acidizing n: the procedure by which acid flow is confined to the natural permeability and porosity of the formation, as opposed to fracture acidizing.

maximum allowable pressure n: the greatest pressure that may safely be applied to a structure, pipe, or vessel. Pressure in excess of this amount leads to failure or explosion.

maximum capacity n: the maximum output of a system or unit (such as a refinery, gasoline plant, pumping unit, or producing well).

maximum efficiency rate n: the producing rate of a well that brings about maximum volumetric recovery from a reservoir with a minimum of residual-oil saturation at the time of depletion. Often maximum efficiency rate (MER) is also used to mean the field production rate that will achieve maximum financial returns from operation of the reservoir. The two rate figures seldom coincide.

maximum water n: in oilwell cementing, the maximum ratio of water to cement that will not cause the water to separate from the slurry on standing.

Mcf abbr: 1,000 cubic feet of gas, commonly used to express the volume of gas produced, transmitted, or consumed in a given period.

Mcf/d abbr: 1,000 ft^3 of gas per day.

md sym: millidarcy.

measure in v: to obtain an accurate measurement of the depth reached in a well by measuring the drill pipe or tubing as it is run into the well.

measuring device n: a special reel and power arrangement for single-stranded wireline to make depth measurements in a well. A calibrated wheel and roller assembly is used to measure the footage of wireline as it is lowered into the well.

measuring tank n: a calibrated tank that, by means of weirs, float switches, pressure switches, or similar devices, automatically measures the volume of liquid run in and then released. Measuring tanks are used in LACT systems. Also called metering tanks or dump tanks.

mechanic n: the crew member that is an all-around handyman for the rig's mechanical components. A mechanic is an optional rig crew member.

mechanical brake n: a brake that is actuated by machinery (such as levers or rods) that is directly linked to it.

mechanical-drive rig n: also called a mechanical rig. See *mechanical rig*.

mechanical jar n: a percussion tool operated mechanically to give an upward thrust to a fish by the sudden release of a tripping device inside the tool. If the fish can be freed by an upward blow, the mechanical jar can be very effective. Also called a hydraulic jar.

mechanical rig n: a drilling rig in which the source of power is one or more internal-combustion engines and in which the power is distributed to rig components through mechanical devices (such as chains, sprockets, clutches, and shafts). Also called a power rig.

megajoule n: the metric unit of service given by a hoisting line in moving 1 000 newtons of load over a distance of 1 000 metres.

MER abbr: maximum efficiency rate.

mercaptan n: a compound chemically similar to alcohol, with sulfur replacing oxygen in the chemical structure. Many mercaptans have an offensive odor and are used as odorants in natural gas.

merchantable oil n: a crude oil in which the BS&W content is not in excess of that allowed for the purchase and therefore salable.

meridian n: a north-south line from which longitudes and azimuths are reckoned.

metacenter n: a point located somewhere on a line drawn vertically through the center of buoyancy of the hull of a floating vessel with the hull in one position (e.g., level) and then another (e.g., inclined). When the hull inclines slightly to a new position, the center of buoyancy of the hull also moves to a new position. If a second line is drawn vertically through the new center of buoyancy, it intersects the first line at a point called the metacenter. Location of the metacenter is impor-

metal - micellar-polymer flooding

tant because it affects the stability of floating vessels (such as mobile offshore drilling rigs).

metal n: opaque crystalline material, usually of high strength, that has good thermal and electrical conductivity, ductility, and reflectivity.

metallic circuit n: the path of electric current through the metallic portions of a corrosion cell.

metamorphic rock n: a rock derived from pre-existing rocks by mineralogical, chemical, and structural alterations caused by processes within the earth's crust. Marble is a metamorphic rock.

meter n: a device used to measure and often record volumes, quantities, or rates of flow of gases, liquids, or electric currents. v: to measure quantities or properties of a substance.

meter calibration n: 1. the operation by which meter readings are compared with an accepted standard. 2. adjustment of a meter so that its readings conform to a standard.

meter chart n: a circular chart of special paper that shows the range of differential pressure and static pressure and is marked by the recording pens of a flow meter.

metering separator n: a complete separator and volume meter integrated into a single vessel. Two-phase units separate oil and gas and meter the oil; three-phase units separate oil, water, and gas, and meter the oil and water.

metering tank n: See *measuring tank*.

meter prover n: a device used to check the accuracy of a meter. Meters are proved by passing a known volume of fluid through them at a known rate and comparing this volume with a standard. A typical prover used for checking meters is the bell prover, which consists essentially of a bell, or piston, that is lowered into a tank containing a known volume of fluid. As the bell is lowered, it displaces the fluid and forces it through the meter to be tested. The rate at which the bell drops is controlled, and thus the rate of flow through the meter can be controlled. The piston prover is extensively used in LACT systems. It has a calibrated tube containing a known volume of fluid. As the piston advances, displacing the fluid in the tube and forcing it through the meter, the volume passing through the meter is recorded. The reading is then compared with the standard (the volume in the tube).

meter run point n: the point in a gas gathering system at which a field measuring meter and accessories are situated.

methane n: a light, gaseous, flammable paraffin hydrocarbon, CH_4, that has a boiling point of $-258°$ F and is the chief component of natural gas and an important basic hydrocarbon for petrochemical manufacture.

methane series n: the paraffin series of hydrocarbons.

methanol (methyl alcohol) n: the lightest alcohol, having the chemical formula CH_3OH. Also called wood alcohol.

metre n: the fundamental unit of length in the metric system. Its symbol is m. It is equal to about 3.28 feet, 39.37 inches, or 100 centimetres.

metric system n: a decimal system of weights and measures based on the metre as the unit of length, the gram as the unit of weight, the cubic metre as the unit of volume, the litre as the unit of capacity, and the square metre as the unit of area.

metric ton n: a measurement equal to 1 000 kg or 2,204.6 lb avoirdupois. In many oil-producing countries, production is reported in metric tons. One metric ton is equivalent to about 7.4 barrels (42 U.S. gal = 1 bbl) of crude oil with a specific gravity of 0.84, or 36° API. In the SI system, it is called a tonne.

mg sym: milligram.

mica n: a silicate mineral characterized by sheet cleavage. Biotite is ferromagnesian black mica, and muscovite is potassic white mica. Sometimes mica is used as a lost circulation material in drilling.

micellar-polymer flooding n: a method of enhanced oil recovery in which chemicals dissolved in water are pumped into a reservoir through injection wells in order to mobilize oil left behind after primary recovery and to move it toward producing wells. The chemical solution includes surfactants or surfactant-forming chemicals that reduce the interfacial and capillary forces between oil and water, dissolve the oil, and carry it out of the pores where it has been trapped. The solution may also contain cosurfactants to match the viscosity of the solution to that of the oil, to stabilize the solution, and to prevent its adsorption by reservoir rock. An electrolyte is often added to aid in adjusting viscosity. Injection of the chemical solution is followed by a slug of water thickened with a polymer, which pushes the dissolved oil through the reservoir, decreases the permeability of established channels so that new channels are opened, and serves as a mobility buffer between the chemical solution and the final injection of water.

Microlog *n:* trade name for a special electric survey method in which three closely spaced electrodes are pressed against the wall of the borehole to obtain a measurement of formation characteristics next to the wall of the hole.

micron *n:* one-millionth of a metre; a metric unit of measure of length equal to 0.001 mm.

microresistivity log *n:* a resistivity logging tool consisting of a spring device and a pad. While the spring device holds the pad firmly against the borehole sidewall, electrodes in the pad measure resistivities in mud cake and nearby formation rock. See *resistivity well logging.*

MICT *abbr:* moving in cable tools; used in drilling reports.

migration *n:* the movement of oil from the area in which it was formed to a reservoir rock where it can accumulate.

mill *n:* a downhole tool with rough, sharp, extremely hard cutting surfaces for removing metal by grinding or cutting. Mills are run on drill pipe or tubing to grind up debris in the hole, remove stuck portions of drill stem or sections of casing for sidetracking, and ream out tight spots in the casing. They are also called junk mills, reaming mills, and so forth, depending on what they do. *v:* to use a mill to cut or grind metal objects that must be removed from a well.

milled bit *n:* also called a milled-tooth bit or a steel-tooth bit. See *steel-tooth bit.*

milled-tooth bit *n:* also called milled bit or steel-tooth bit. See *steel-tooth bit.*

millidarcy *n:* one-thousandth of a darcy.

millilitre *n:* one-thousandth of a litre. In analyzing drilling mud, this term is used interchangeably with cubic centimetre (cm^3). A quart equals 964 ml.

millimetre *n:* a measurement unit in the metric system equal to 10^{-3} metre (0.001 metre). Its symbol is mm, and it is used to measure pipe and bit diameter, nozzle size, liner length and diameter, and cake thickness.

milling tool *n:* the tool used in the operation of milling. See *mill.*

millisec *abbr:* millisecond.

mill scale *n:* thin, dense oxide scale that forms on the surface of newly manufactured steel as the steel cools. Mill scale can become cathodic to its own steel base, forming galvanic corrosion cells.

min *abbr:* minute.

mineral rights *n:* the rights of ownership, conveyed by deed, of gas, oil, and other minerals beneath the surface of the earth. In the United States, mineral rights are the property of the surface owner unless disposed of separately.

Mine Safety and Health Administration *n:* a U.S. government agency that evaluates research in the causes of occupational diseases and accidents. Headquartered in Arlington, Virginia, MSHA is responsible for administration of the certification of respiratory safety equipment.

miniaturized completion *n:* a well completion in which the production casing is less than 4.5 inches in diameter.

minilog *n:* See *pad resistivity instrument.*

MIR *abbr:* moving in rig; used in drilling reports.

miscible *adj:* capable of being mixed; capable of mixing in any ratio without separation of the two phases.

miscible drive *n:* a method of enhanced recovery in which various hydrocarbon solvents or gases (such as propane, LPG, natural gas, carbon dioxide, or a mixture thereof) are injected into the reservoir to reduce interfacial forces between oil and water in the pore channels and thus displace oil from the reservoir rock.

miscible flood *n:* a method of secondary recovery of fluids from a reservoir by injection of fluids that are miscible with the reservoir fluids.

mist drilling *n:* a drilling technique that uses air or gas to which a foaming agent has been added.

mist extractor *n:* a metal member used to remove small droplets of moisture or condensable hydrocarbons from a gas stream in an oil and gas separator. The small droplets of moisture collect on the metal surface to form larger drops, which are removed from the separator along with other separated liquids.

mixed butane *n:* also called field-grade butane. See *field-grade butane.*

mixed-gas diving *n:* diving in which a diver uses a breathing medium of oxygen and one or more inert gases synthetically mixed.

mixed string *n:* a combination string. See *casing string.*

mixing tank *n:* any tank or vessel used to mix components of a substance (as in the mixing of additives with drilling mud).

mix mud *v:* to prepare drilling fluids from a mixture of water or other liquids and one or more of the various dry mud-making materials (such as clay, weighting materials, chemicals, etc.).

ml *sym:* millilitre.

mm *sym:* millimetre.

mm² *sym:* square millimetre.

mm³ *sym:* cubic millimetre.

MMcf *abbr:* million cubic feet; a common unit of measurement for large quantities of gas.

MMscf *abbr:* million standard cubic feet. The standard referred to is usually 60° F and 1 atmosphere (14.7 psi) of pressure but varies from state to state.

MMscf/d *abbr:* million standard cubic feet per day.

MO *abbr:* moving out; used in drilling reports.

mobile offshore drilling rig *n:* a drilling rig that is used exclusively to drill offshore wells and that floats upon the surface of the water when being moved from one location to another. It may or may not float once drilling begins. The drill ship, semisubmersible drilling rig, and jackup drilling rig are all mobile rigs; a platform rig is not.

modified cement *n:* a cement whose properties, chemical or physical, have been altered by additives.

modular-spaced workover rig *n:* an offshore platform rig designed in equipment packages or modules that are light enough to be lifted onto a platform by a platform crane. In most cases, the maximum weight of a module is 12,000 pounds. Once lifted from the work boat, the rig can be erected and working within twenty-four to thirty-six hours. As in all mast-type rigs, working depth is limited by the strength of the mast, typically 12,000 to 14,000 feet.

mol *sym:* mole.

mole *n:* the fundamental unit of mass of a substance. Its symbol is mol. A mole of any substance is the number of grams or pounds indicated by its molecular weight. For example, water, H_2O, has a molecular weight of approximately 18. Therefore, a gram mole of water is 18 grams of water; a pound mole of water is 18 pounds of water. See *molecular weight*.

molecular sieves *n pl:* synthetic zeolites packaged in bead or pellet form for (1) use in recovering contaminants or impurities from liquid and vapor product streams by selective adsorption and for (2) use as a catalyst.

molecular weight *n:* the sum of the atomic weights in a molecule. For example, the molecular weight of water, H_2O, is 18 because the atomic weight of each of the hydrogen molecules is 1 and the atomic weight of oxygen is 16. See *mole*.

molecule *n:* the smallest part of a compound that can exist on its own. The atoms of which it consists may be different (such as the hydrogen and oxygen atoms of water, H_2O) or the same (such as the two hydrogen atoms of free hydrogen, H_2). See *atom* and *compound*.

mole percent *n:* the ratio of the number of moles of one substance to the total number of moles in a mixture of substances, multiplied by 100 (to put the number on a percentage basis).

moment *n:* a turning effect created by a force F acting at a perpendicular distance S from the center of rotation; the product of a force and a distance to a particular axis or point.

Monel steel *n:* a nickel-base alloy containing copper, iron, manganese, silicon, and carbon. Nonmagnetic drill collars are often made of this material.

monitor *n:* an instrument that reports the performance of a control device or signals if unusual conditions appear in a system. For example, a BS&W monitor provides a mechanical means of preventing contaminated oil from entering the pipeline by detecting the presence of excessive water and actuating valves to divert the flow back to dehydration facilities.

monkeyboard *n:* the derrickman's working platform. As pipe or tubing is run into or out of the hole, the derrickman must handle the top end of the pipe, which may be as high as 90 feet (27 m) in the derrick or mast. The monkeyboard provides a small platform to raise him to the proper height for handling the top of the pipe.

monocline *n:* rock strata that dip in one direction only. Compare *anticline* and *syncline*.

montmorillonite *n:* a clay mineral often used as an additive to drilling mud. It is a hydrous aluminum silicate capable of reacting with such substances as magnesium and calcium. See *bentonite*.

moon pool *n:* a walled round hole or well in the hull of a drill ship (usually in the center) through which the drilling assembly and other assemblies pass while a well is being drilled, completed, or abandoned from the drill ship.

morning tour *n:* (pronounced "tower") usually called daylight tour. See *daylight tour*.

mosquito bill *n:* a tube mounted at the bottom of a sucker rod pump and inside a gas anchor to provide a conduit into the pump for well fluids that contain little or no gas.

mother hubbard *n:* (slang) also called mud box or mud saver. See *mud box*.

motion compensator *n:* any device (such as a bumper sub or heave compensator) that serves to maintain constant weight on the bit in spite of vertical motion of a floating offshore drilling rig.

motor *n:* a hydraulic, air, or electric device used to do work. A motor is not an engine.

motor-generator rig *n:* a drilling rig driven by electric motors with current supplied by engine-driven generators at the rig.

motorman *n:* the crew member on a rotary drilling rig who is responsible for the care and operation of drilling engines.

motor valve *n:* a valve operated by power other than manual (i.e., hydraulic, electric, or mechanical).

mousehole *n:* an opening through the rig floor, usually lined with pipe, into which a length of drill pipe is placed temporarily for later connection to the drill string.

mousehole connection *n:* the procedure of adding a length of drill pipe or tubing to the active string, in which the length to be added is placed in the mousehole, made up to the kelly, then pulled out of the mousehole, and subsequently made up into the string.

mousetrap *n:* a fishing tool used to recover a parted string of sucker rods from a well.

MPa *sym:* megapascal.

mph *abbr:* miles per hour.

µs *sym:* microsecond.

Mscf/D *abbr:* thousand standard cubic feet per day.

MSHA *abbr:* Mine Safety and Health Administration.

mud *n:* the liquid circulated through the wellbore during rotary drilling and workover operations. In addition to its function of bringing cuttings to the surface, drilling mud cools and lubricates the bit and drill stem, protects against blowouts by holding back subsurface pressures, and deposits a mud cake on the wall of the borehole to prevent loss of fluids to the formation. Although it was originally a suspension of earth solids (especially clays) in water, the mud used in modern drilling operations is a more complex, three-phase mixture of liquids, reactive solids, and inert solids. The liquid phase may be fresh water, diesel oil, or crude oil and may contain one or more conditioners. See *drilling fluid*.

mud acid *n:* a mixture of hydrochloric and hydrofluoric acids and surfactants, used to effect mud removal from the wellbore.

mud additive *n:* any material added to drilling fluid to change some of its characteristics or properties.

mud analysis *n:* examination and testing of drilling mud to determine its physical and chemical properties.

mud balance *n:* a beam balance consisting of a cup and a graduated arm carrying a sliding weight and resting on a fulcrum, used to determine the density or weight of drilling mud.

mud box *n:* a hinged, cylindrical metal device placed around a joint of pipe as it is being broken out during a trip out of the hole. It keeps mud from splashing beyond the immediate area. It is also called a mud saver, a splash box, a wet box, or a mother hubbard.

mud cake *n:* the sheath of mud solids that forms on the wall of the hole when liquid from mud filters into the formation; also called wall cake or filter cake.

mud circulation *n:* the process of pumping mud downward to the bit and back up to the surface in a drilling or workover operation. See *normal circulation* and *reverse circulation*.

mud conditioning *n:* the treatment and control of drilling mud to ensure that it has the correct properties. Conditioning may include the use of additives, the removal of sand or other solids, the removal of gas, the addition of water, and other measures to prepare the mud for conditions encountered in a specific well.

mud density recorder *n:* a device that automatically records the weight or density of drilling fluid as it is being circulated in a well.

mud engineer *n:* a person whose duty is to test and maintain the properties of the drilling mud that are specified by the operator.

mud-flow indicator *n:* a device that continually measures and may record the flow rate of mud returning from the annulus and flowing out of the mud return line. If the mud does not flow at a fairly constant rate, a kick or lost circulation may have occurred.

mud-flow sensor n: also called mud-flow indicator. See *mud-flow indicator*.

mud-gas separator n: a device that separates gas from the mud coming out of a well when gas cutting has occurred or when a kick is being circulated out.

mud gun n: a pipe that shoots a jet of drilling mud under high pressure into the mud pit to mix additives with the mud or to agitate the mud.

mud hopper n: See *hopper*.

mud hose n: also called kelly hose or rotary hose. See *rotary hose*.

Mud-kil n: trade name for a chemical additive for portland cement that reduces the effect of contamination of cementing slurries by the organic chemicals commonly found in drilling muds.

mud-level recorder n: a device that measures and records the height (level) of the drilling fluid in the mud pits. The level of the mud in the pits should remain fairly constant during the drilling of a well. However, if the level rises, then the possibility of a kick or a blowout exists. Conversely, if the level falls, then loss of circulation may have occurred.

mud log n: a record of information derived from examination of drilling fluid and drill bit cuttings. See *mud logging*.

mud logger n: an employee of a mud logging company who performs mud logging.

mud logging n: the recording of information derived from examination and analysis of formation cuttings made by the bit and of mud circulated out of the hole. A portion of the mud is diverted through a gas-detecting device. Cuttings brought up by the mud are examined under ultraviolet light to detect the presence of oil or gas. Mud logging is often carried out in a portable laboratory set up at the well.

mud man n: also called a mud engineer. See *mud engineer*.

mud motor n: See *Dyna-Drill* and *turbodrill*.

mud-off v: 1. to seal the hole against formation fluids by allowing the buildup of wall cake. 2. to block off the flow of oil into the wellbore.

mud pit n: an open pit dug in the ground to hold drilling fluid or waste materials discarded after the treatment of drilling mud. For some drilling operations, mud pits are used for suction to the mud pumps, settling of mud sediments, and storage of reserve mud. Steel tanks are much more commonly used for these purposes now, but they are still sometimes referred to as pits.

mud pump n: a large, high-pressure reciprocating pump used to circulate the mud on a drilling rig. A typical mud pump is a two-cylinder, double-acting or a three-cylinder, single-acting piston pump whose pistons travel in replaceable liners and are driven by a crankshaft actuated by an engine or a motor. Also called a slush pump.

mud report n: a special form that is filled out by the mud man and that records the properties of the drilling mud used while a well is being drilled.

mud return line n: a trough or pipe that is placed between the surface connections at the wellbore and the shale shaker and through which drilling mud flows upon its return to the surface from the hole. Also called flow line.

mud saver n: also called mud box or mother hubbard. See *mud box*.

mud screen n: also called a shale shaker. See *shale shaker*.

mud seal n: a closing device that prevents the entrance of mud.

mud suction pit n: See *suction pit*.

mud tank n: one of a series of open tanks, usually made of steel plate, through which the drilling mud is cycled to allow sand and fine sediments to be removed. Additives are mixed with the mud in the tanks, and the fluid is temporarily stored there before being pumped back into the well. Modern rotary drilling rigs are generally provided with three or more tanks, fitted with built-in piping, valves, and mud agitators. Also called mud pits.

mud-up v: to add solid materials (such as bentonite or other clay) to a drilling fluid composed mainly of clear water to obtain certain desirable properties.

mud weight n: a measure of the density of a drilling fluid expressed as pounds per gallon (ppg), pounds per cubic foot (lb/ft^3), or kilograms per cubic metre (kg/m^3). Mud weight is directly related to the amount of pressure the column of drilling mud exerts at the bottom of the hole.

mud weight recorder n: an instrument, installed in the mud pits, that has a recorder mounted on the rig floor to provide a continuous reading of the mud weight.

muffler n: a device used to reduce exhaust noise to an acceptable level.

mule shoe n: a sub, shaped like a horseshoe, that is used to orient the drill stem downhole. See *sub*.

multiple completion n: an arrangement for producing a well in which one wellbore penetrates two or more petroleum-bearing formations that lie one over the other. The tubing strings are suspended side by side in the production casing string, each a different length and each packed off to prevent the commingling of different reservoir fluids. Each reservoir is then produced through its own tubing string.

multiple well pumping system *n:* a method of lifting oil out of several wells in a field. A pump is placed at every well; however, all the pumps are powered by a single prime mover (engine or motor) instead of each pump being powered individually.

multishot survey *n:* a directional survey that provides many records of the hole. See *directional survey*.

multistage cementing tool *n:* a device used for cementing two or more separate sections behind a casing string, usually for a long column that might cause formation breakdown if the cement were displaced from the bottom of the string.

muriatic acid *n:* hydrochloric acid.

mV *sym:* millivolt.

N

N *sym:* newton.

NACE *abbr:* National Association of Corrosion Engineers.

naphtha *n:* a volatile, flammable liquid hydrocarbon distilled from petroleum and used as a solvent or a fuel.

naphthene-base oil *n:* a crude oil that is characterized by a low API gravity and a low yield of lubricating oils and that has a low pour point and a low viscosity index (compared to paraffin-base oils). It is often called asphalt-base oil because the residue from its distillation contains asphaltic materials but little or no paraffin wax.

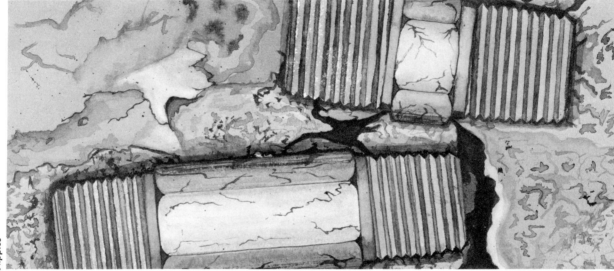

Nipples

naphthene series *n:* the saturated hydrocarbon compounds of the general formula C_nH_{2n} (e.g., ethylene or ethene, C_2H_4). See *hydrocarbons*.

National Association of Corrosion Engineers *n:* organization whose function is to establish standards and recommended practices for the field of corrosion control. It is based in Houston, Texas.

National Institute of Occupational Safety and Health *n:* a U.S. government agency that conducts research in the causes of occupational diseases and accidents. Headquartered in Rockville, Maryland, NIOSH is responsible for administration of the certification of respiratory safety equipment.

National LP-Gas Association *n:* an association whose members are producers or marketers of butane-propane gas and manufacturers of equipment and appliances. The NLPGA develops standards, provides safety programs, conducts market research and develops industry statistics, and provides government liaison on laws and regulations.

natural gas *n:* a highly compressible, highly expansible mixture of hydrocarbons having a low specific gravity and occurring naturally in gaseous form. Besides hydrocarbon gases, natural gas may contain appreciable quantities of nitrogen, helium, carbon dioxide, hydrogen sulfide, and water vapor. Although gaseous at normal temperatures and pressures, the gases comprising the mixture that is natural gas are variable in form and may be found either as gases or as liquids under suitable conditions of temperature and pressure.

natural gas liquids *n:* those hydrocarbons liquefied at the surface in field facilities or in gas processing plants. Natural gas liquids include propane, butane, and natural gasoline.

natural gasoline n: the liquid hydrocarbons recovered from wet natural gas; casinghead gasoline. See *casinghead gasoline*.

natural gas plant n: also called natural gas processing plant. See *natural gas processing plant*.

natural gas processing plant n: an installation in which natural gas is processed for recovery of natural gas liquids, the heavier hydrocarbon components of natural gas, including liquefied petroleum gases (such as butane and propane) and natural gasoline. The modern and preferred term for gas processing plant, natural gas plant, gasoline plant, and others.

naturally aspirated adj: term used to describe a diesel engine in which air flows into the engine by means of atmospheric pressure only.

neat cement n: a cement with no additives other than water.

neck down v: to taper to a reduced diameter. A pipe becomes necked down when it is subjected to excessive longitudinal stress.

necking n: the tendency of a metal bar or pipe to taper to a reduced diameter at some point when subjected to excessive longitudinal stress. See *bottleneck*.

needle valve n: a form of globe valve that contains a sharp-pointed, needlelike plug that is driven into and out of a cone-shaped seat to accurately control a relatively small rate of flow of a fluid. In a fuel injector, the fuel pressure forces the needle valve off its seat to allow injection to take place.

net observed volume n: the total volume of all petroleum liquids, excluding sediment and water and free water, at observed temperature and pressure.

net-oil computer n: a system of electronic and mechanical devices that automatically determines the amount of oil in a water and oil emulsion. One advantage of a net-oil computer is that the water and oil do not have to be separated for measuring the volume of the oil.

net production n: the amount of oil produced by a well or a lease, exclusive of its BS&W content. Net production is also called working-interest oil (i.e., the net oil produced by all of its wells multiplied by the working interest of a company in the wells).

net standard volume n: the total volume of all petroleum liquids, excluding sediment and water and free water, corrected by the appropriate temperature correction factor (D_{t1}) for the observed temperature and API gravity, relative density, or density to a standard temperature such as 60°F or 15°C, and also corrected by the applicable pressure correction factor (C_{p1}) and meter factor.

net tonnage n: the gross tonnage of a ship or a mobile offshore drilling rig less all spaces that are not or cannot be used for carrying cargo, expressed in tons equal to 100 cubic feet.

neutrino n: a neutral particle ejected from the nucleus of an atom when the neutron-to-proton ratio is too high. Neutrinos are difficult to detect because of their penetrating power and are not recorded in logging.

neutron n: a part of the nucleus of all atoms except hydrogen. Under certain conditions neutrons can be emitted from a substance when its nucleus is penetrated by gamma particles from a highly radioactive source. This phenomenon is used in neutron logging.

neutron-epithermal neutron log n: a log designed to have maximum sensitivity to detectable neutrons with energies above the thermal level and minimal sensitivity to capture gamma rays and thermal neutrons. Since the neutron-epithermal neutron log is only slightly affected by capture cross-section and capture gamma ray emission, it reduces errors that arise from variations in formation chemistry.

neutron-gamma log n: a record measuring the capture gamma rays emitted by a nucleus when it captures a neutron in a collision.

Neutron Lifetime Log n: trade name for a pulsed-neutron survey.

neutron log n: See *radioactivity logging*.

neutron radiation n: radiation produced by nuclear disintegration, as when a nucleus is penetrated by gamma rays from a highly radioactive source. Neutron radiation can penetrate several feet of lead and is so difficult to observe that it remained undiscovered long after alpha, beta, and gamma rays were well known.

neutron-thermal neutron log n: a record predominantly sensitive to thermal neutrons. Thermal detectors have a varying sensitivity to gamma rays.

newton n: the unit of force in the metric system; its symbol is N. A newton is the force required to accelerate an object of 1 kilogram mass to a velocity of 1 metre per second in 1 second.

Newtonian fluid n: a fluid in which the viscosity remains constant for all rates of shear if constant conditions of temperature and pressure are maintained. Most drilling fluids behave as non-Newtonian fluids, as their viscosity is not constant but varies with the rate of shear.

newton-metre n: also called a joule. See *joule*.

NGL abbr: natural gas liquids.

NIOSH abbr: National Institute of Occupational Safety and Health.

nipple n: a short, threaded tubular coupling, used for making connections between pipe joints and other tools.

nipple chaser n: (slang) a crew member who procures and delivers tools and equipment for a drilling rig.

nipple up v: in drilling, to assemble the blowout preventer stack on the wellhead at the surface.

nitrogen narcosis n: the intoxicating or narcotic effect of gaseous nitrogen, experienced by a diver breathing air below approximately 100 feet of depth. The effect increases with depth, impairing a diver's ability to think and act effectively.

nitro shooting n: a formation-stimulation process first used about a hundred years ago in Pennsylvania. Nitroglycerine is placed in a well and exploded to fracture the rock. Sand and gravel or cement is usually placed above the explosive charge to improve the efficiency of the shot. Today nitro shooting has been largely replaced by formation fracturing.

NLPGA abbr: National LP-Gas Association

noble metal n: any of the metals with low reactive tendencies at the upper end of the electrochemical series.

nominal size n: a designated size that may be different from the actual size.

nomograph n: a chart that represents an equation containing a number of variables in the form of scales so that a straight line cuts the scales at values of the variables satisfying the equation.

nonane n: a paraffin hydrocarbon, C_9H_{20}, that is liquid at atmospheric conditions. Its boiling point is about 303.5° F (at 14.7 psi).

nonferrous alloy n: alloy containing less than 50 percent iron.

nonmagnetic drill collar n: a drill collar made of an alloy that does not affect the readings of a magnetic compass placed within it to obtain subsurface indications of the direction of a deviated wellbore. Used in directional drilling.

nonporous adj: containing no interstices; having no pores.

normal butane n: in commercial transactions, a product meeting GPA specification for commercial butane and, in addition, containing a minimum of 95 liquid volume percent normal butane. Chemically, normal butane is an aliphatic compound of the paraffin series having the chemical formula C_4H_{10} and having all its carbon atoms joined in a straight chain.

normal circulation n: the smooth, uninterrupted circulation of drilling fluid down the drill stem, out the bit, up the annular space between the pipe and the hole, and back to the surface.

normal formation pressure n: formation fluid pressure equivalent to 0.465 psi per foot of depth from the surface. If the formation pressure is 4,650 psi at 10,000 feet, it is considered normal.

normalizing n: heat-treating applied to metal tubular goods to ensure uniformity of the grain structure of the metal.

nose button n: a hard-metal projection that is placed on the end of the pilot pin of a roller cone bit and that serves to absorb some of the wear created by outward thrusts as the bit rotates.

notch fatigue n: metal fatigue concentrated by surface imperfection, either mechanical (such as a notch) or metallurgical (defect in the metal itself).

nozzle n: 1. a passageway through jet bits that allows the drilling fluid to reach the bottom of the hole and flush the cuttings through the annulus. Nozzles come in different sizes that can be interchanged on the bit to allow more or less flow. 2. the part of the fuel system of an engine that has small holes in it to permit fuel to enter the cylinder. Properly known as a fuel-injection nozzle. Also called a spray valve. The needle valve is directly above the nozzle.

NS abbr: no show; used in drilling reports.

nutating meter n: a flow meter that operates on the principle of the positive displacement of fluid by incorporating the wobbling motion of a piston or a disc. See *positive-displacement meter*.

O&G *abbr:* oil and gas; used in drilling reports.

O&GCM *abbr:* oil- and gas-cut mud; used in drilling reports.

O&SW *abbr:* oil and salt water; used in drilling reports.

OBQ *abbr:* on-board quantity.

OC *abbr:* oil-cut; used in drilling reports.

Occupational Safety and Health Administration *n:* a U.S. government agency that establishes and enforces safety standards for industry employees.

OCM *abbr:* oil-cut mud; used in drilling reports.

OCS *abbr:* outer continental shelf.

OCS orders *n:* rules and regulations, set by the U.S. Geological Survey, that govern oil operations in U.S. waters on the outer continental shelf.

octane *n:* a paraffin hydrocarbon, C_8H_{18}, that is a liquid at atmospheric conditions. Its boiling point is 258°F (at 14.7 psi).

octane rating *n:* a classification of gasoline according to its antiknock qualities. The higher the octane number, or rating, the greater are the antiknock qualities of the gasoline.

OD *abbr:* outside diameter.

odorant *n:* a chemical, usually a mercaptan, that is added to natural gas so that the presence of the gas can be detected by the smell.

OF *abbr:* open flow; used in drilling reports.

off-production *adj:* shut in or temporarily unable to produce (said of a well).

offset link *n:* in transmission chain, a combination of roller link and pin link used when a chain has an odd number of pitches.

offset well *n:* a well drilled on a tract of land next to another owner's tract on which there is a producing well.

offset-well data *n:* information obtained from wells that are drilled in an area close to where a well is being drilled or worked over. Such information can be very helpful in determining how a particular well will behave or react to certain treatments or techniques applied to it.

offshore *n:* that geographic area which lies seaward of the coastline. In general, the term *coastline* means the line of ordinary low water along that portion of the coast that is in direct contact with the open sea or the line marking the seaward limit of inland waters.

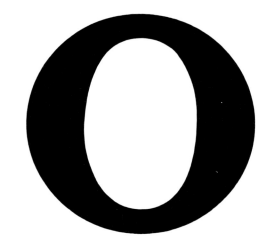

Orifice meter

offshore drilling n: drilling for oil in an ocean, gulf, or sea, usually on the continental shelf. A drilling unit for offshore operations may be a mobile floating vessel with a ship or barge hull, a semisubmersible or submersible base, a self-propelled or towed structure with jacking legs (jackup drilling rig), or a permanent structure used as a production platform when drilling is completed. In general, wildcat wells are drilled from mobile floating vessels or from jackups, while development wells are drilled from platforms.

offshore rig n: any of various types of drilling structures designed for use in drilling wells in oceans, seas, bays, gulfs, and so forth. Offshore rigs include platforms, jackup drilling rigs, semisubmersible drilling rigs, submersible drilling rigs, and drill ships.

OH abbr: open hole; used in drilling reports.

ohmmeter n: an instrument for measuring electric resistance in ohms.

ohm-metre n: a unit for measuring electrical resistance. If a container with sides of 1 metre each is filled with a solution and a resistance of 1 ohm is measured when a current is passed through the container from one face to the opposite face, the resistivity of the unit volume is 1 ohm-metre.

Ohm's law n: a law that concerns the behavior of electrical flow through a conductor. Ohm's law is stated as

$$R = E/I$$

where

R = resistance, E = volts, and I = current.

The law is used in measuring the resistivity of a substance to the flow of electric current.

oil and gas separator n: an item of production equipment used to separate liquid components of the well stream from gaseous elements. Separators are either vertical or horizontal and either cylindrical or spherical in shape. Separation is accomplished principally by gravity, the heavier liquids falling to the bottom and the gas rising to the top. A float valve or other liquid-level control regulates the level of oil in the bottom of the separator.

oil-base mud n: an oil that contains from less than 2 percent up to 5 percent water. The water is spread out, or dispersed, in the oil as small droplets. See *oil mud* and *invert-emulsion mud*.

oil-emulsion mud n: a water-base mud in which water is the continuous phase and oil is the dispersed phase. The oil is spread out, or dispersed, in the water in small droplets, which are tightly emulsified so that they do not settle out. Because of its lubricating abilities, an oil-emulsion mud increases the drilling rate and ensures better hole conditions than other muds. Compare *oil mud*.

oil field n: the surface area overlying an oil reservoir or reservoirs. Commonly, the term includes not only the surface area, but also the reservoir, the wells, and the production equipment.

oil in place n: crude oil that is estimated to exist in a reservoir but has not been produced.

oil mud n: a drilling mud in which oil is the continuous phase. Oil-base mud and invert-emulsion mud are types of oil muds. They are useful in drilling certain formations that may be difficult or costly to drill with water-base mud. Compare *oil-emulsion mud*.

oil operator n: also called operator. See *operator*.

oil patch n: (slang) the oil field.

oil ring n: the ring or rings that are located on the lower portion of a piston. They prevent excessive oil from being drawn into the combustion space during the suction stroke.

oil sand n: 1. a sandstone that yields oil. 2. (by extension) any reservoir that yields oil, whether or not it is sandstone.

oil saver n: a gland arrangement that seals by pressure and is used to prevent leakage and waste of gas, oil, or water around a wireline (as when swabbing a well). It is operated either mechanically or hydraulically.

oil scout n: a representative of an oil company who gathers data on new oil and gas wells and other industry developments.

oil seep n: a surface location where oil appears, the oil having permeated its subsurface boundaries and accumulated in small pools.

oil shale n: a formation containing hydrocarbons that cannot be recovered by an ordinary oilwell but can be mined. After processing, the hydrocarbons are extracted from the shale. The cost of mining and treatment of oil shale has until recently been too great to compete with the cost of oilwell drilling.

oil slick n: a film of oil floating on water, considered a pollutant.

oil spill n: a quantity of oil that has leaked or fallen onto the ground or onto the surface of a body of water.

oil string n: the final string of casing set in a well after the productive capacity of the formation has been determined to be sufficient; the long string or production casing.

oil-water contact *n:* the point or plane at which the bottom of an oil sand contacts the top of a water sand in a reservoir; the oil-water interface.

oil-water emulsion *n:* See *reverse emulsion* and *emulsion*.

oilwell cement *n:* cement or a mixture of cement and other materials for use in oil, gas, or water wells.

oilwell pump *n:* any pump, surface or subsurface, that is used to lift fluids from the reservoir to the surface. See *sucker rod pumping* and *hydraulic pumping*.

oil-wet rock *n:* See *wettability*.

oil zone *n:* a formation or horizon of a well from which oil may be produced. The oil zone is usually immediately under the gas zone and on top of the water zone if all three fluids are present and segregated.

olefins units *n pl:* the units in a refinery that produce ethylene and propylene, both for the polymer units (to make polyethylene and polypropylene) and for sale to other chemical and plastic producers.

on-board quantity *n:* the measurable or estimatable materials remaining on board in vessel cargo tanks and pipelines prior to loading. Includes water, oil, slops, oily residue, oil/water emulsion, sludge, and sediment. Its abbreviation is OBQ.

on-deck *adj:* present on a ship or rig deck and exposed to weather.

one-step grooving system *n:* a pattern of drum spooling in which the wire rope is controlled by grooves to move parallel to drum flanges for 70-80 percent of the circumference and then crosses over to start the next wrap.

on-stream *adj:* 1. of a pump or pump station, moving oil by pumping. 2. of a gas processing plant or refinery, in operation or running.

on-suction *adj:* of a tank, open to pump suction.

on-the-line *adj:* of a tank, being emptied into a pipeline.

on-the-pump *adj:* of a well, being pumped.

ool *abbr:* oolitic; used in drilling reports.

OPEC *abbr:* Organization of Petroleum Exporting Countries.

open circuit *n:* a diving life-support system in which the diver's exhalation is vented completely to the water.

open-circuit regulator *n:* also called demand regulator. See *demand regulator*.

open flow potential *n:* the theoretical maximum capacity of a gas well as determined by a test conducted under limiting conditions. The method of determining this potential varies from state to state.

open flow test *n:* a test made to determine the volume of gas that will flow from a well during a given time span when all surface control valves are wide open.

open formation *n:* a petroleum-bearing rock with good porosity and permeability.

open hole *n:* 1. any wellbore in which casing has not been set. 2. open or cased hole in which no drill pipe or tubing is suspended. 3. the portion of the wellbore that has no casing.

open-hole completion *n:* a method of preparing a well for production in which no production casing or liner is set opposite the producing formation. Reservoir fluids flow unrestricted into the open wellbore. An open-hole completion has limited use in rather special situations. Also called a barefoot completion.

operating company *n:* also called operator. See *operator*.

operator *n:* the person or company, either proprietor or lessee, actually operating an oilwell or lease. Generally, the oil company by whom the drilling contractor is engaged. Compare *unit operator*.

optimum rate of flow *n:* that rate of flow of fluid from a well that will provide maximum ultimate recovery of fluid from the reservoir.

optimum water *n:* the amount of water used to give a cement slurry the best properties for its particular application.

ore *n:* a mineral from which a valuable substance such as a metal can be extracted.

organic compounds *n pl:* chemical compounds that contain carbon atoms, either in straight chains or in rings, and hydrogen atoms. They may also contain oxygen, nitrogen, or other atoms.

organic rock *n:* rock materials produced by plant or animal life (coal, petroleum, limestone, etc.).

organic theory *n:* an explanation of the origin of petroleum, which holds that the hydrogen and the carbon that make up petroleum come from plants and animals of land and sea. Furthermore, the theory holds that more of this organic material comes from very tiny creatures of swamp and sea than comes from larger creatures of land.

Organization of Petroleum Exporting Countries *n:* an organization of the countries of the Middle East, Africa, and South America that produce oil and export it. The purpose of the organization is to negotiate and regulate oil prices.

orientation *n:* the process of positioning a deflection tool so that it faces in the direction necessary to achieve the desired direction and drift angle for a directional hole.

oriented core *n:* a core sample whose location in the reservoir has been pinpointed.

oriented drill pipe *n:* drill pipe run in a well in a definite position, often a requisite in directional drilling.

orifice *n:* an opening of a measured diameter, used for measuring the flow of fluid through a pipe or delivering a given amount of fluid through a fuel nozzle. In measuring the flow of fluid through a pipe, the orifice must be of smaller diameter than the pipe diameter; it is placed in an orifice plate held by an orifice fitting.

orifice fitting *n:* a device placed in a gas line to hold an orifice plate.

orifice-flange tap *n:* a tap for pressure connections made close to the orifice and through the flanges.

orifice meter *n:* an instrument used to measure the flow of fluid through a pipe. The orifice meter is an inferential device, which measures and records the pressure differential created by the passage of a fluid through an orifice of critical diameter placed in the line. The rate of flow is calculated from the differential pressure and the static, or line, pressure and other factors such as the temperature and density of the fluid, the size of the pipe, and the size of the orifice.

orifice pipe tap *n:* a tap for pressure connections made at points two-and-a-half pipe diameters upstream and eight diameters downstream from the orifice; a full-flow connection.

orifice plate *n:* a sheet of metal, usually circular, in which a hole of specific size is made for use in an orifice fitting.

orifice pressure drop *n:* the pressure differential that occurs across an orifice plate.

orifice well tester *n:* a device used to measure the gas flow from a well, including orifice plates, a hose, and a manometer; static pressure differences before and after a sharp-edged orifice are converted to flow values. It is used primarily for estimating the amount of gas flowing during a drill stem test, when a high degree of accuracy is not required.

oscilloscope *n:* a test instrument that visually records an electrical wave on a fluorescent screen.

OSHA *abbr:* Occupational Safety and Health Administration.

Ouija board *n:* a device used by engineers to determine which way to turn deflection tools in order to get the desired drift angle in a hole. This device is composed of a protractor and various straight-edged scales that pivot around the protractor in order to figure the drift angle.

outage *n:* also called ullage. See *ullage*.

outage gauge *n:* a measure of the volume of liquid in a storage tank from a reference point at the top of the tank to the surface of the liquid.

outboard *adv:* away from the center of the hull or toward the side of an offshore drilling rig.

outcrop *n:* the exposed portion of a buried layer of rock. *v:* to appear on the earth's surface (as a rock).

outer continental shelf orders *n:* See *OCS orders*.

out-of-gauge bit *n:* a bit that is no longer of the proper diameter.

out-of-gauge hole *n:* a hole that is not in gauge – that is, of a size smaller or larger than the diameter of the bit used to drill the hole.

outpost well *n:* a well located outside the established limits of a reservoir; a step-out well.

output shaft *n:* the transmission shaft nearest to the machinery to be driven – driven itself by the shaft driven by the power source.

outrigger *n:* a projecting member run out at an angle from the sides of a portable mast or a land crane to the ground to provide stability and to minimize the possibility of having the mast or the crane overturn.

outside cutter *n:* See *external cutter*.

outside diameter *n:* the distance across the exterior circle, especially in the measurement of pipe. See *diameter*.

outward axial thrust *n:* an outward force created along a centerline drawn through a cone (the axis) as a roller cone bit rotates.

over and short station *n:* a pump station where one or more tanks are floating on the line. See *floating tank*.

overburden *n:* the strata of rock that lie above the stratum of interest in drilling.

overburden pressure *n:* the pressure exerted by the overburden on the formation targeted for drilling.

overflow *n:* the effluent of a cone-shaped centrifuge, passing up the inside of the cone and leaving through the vortex finder.

overflow pipe *n:* a pipe installed at the top of a tank to enable the liquid within it to be discharged to another vessel when the tank is filled to capacity.

overflush *n:* an excess quantity of fluid used to push acid out of the tubing or casing when an acid mixture is put into a well, thus directing the acid to the desired place in the well.

overgauge hole *n:* a hole whose diameter is larger than the diameter of the bit used to drill it. An overgauge hole can occur when a bit is not properly stabilized or does not have enough weight put on it.

overrunning clutch *n:* 1. a special clutch that permits a rotating member to turn freely under certain conditions but not under others. 2. a clutch that is used in a starter and transmits cranking effort but overruns freely when the engine tries to drive the starter.

overshot *n:* a fishing tool that is attached to tubing or drill pipe and lowered over the outside wall of pipe or sucker rods lost or stuck in the wellbore. A friction device in the overshot, usually either a basket or a spiral grapple, firmly grips the pipe, allowing the lost fish to be pulled from the hole.

oxidation *n:* a chemical reaction in which a compound loses electrons and gains a more positive charge.

oxide *n:* a chemical compound in which oxygen is joined with a metal or a nonmetal.

oxyacetylene welding *n:* See *acetylene welding.*

oxygen-concentration cell *n:* a corrosion cell formed by differing concentrations of oxygen in an electrolyte.

oxygen toxicity *n:* a harmful reaction experienced by divers breathing extremely high partial pressures of oxygen. Divers may suffer from two different forms; one affects the central nervous system and another affects the pulmonary muscles. These dangers eliminate the use of pure oxygen as a breathing medium below 50 feet in commercial operations.

oz *abbr:* ounce.

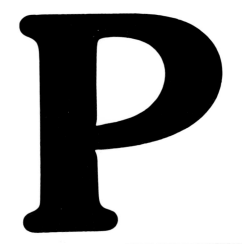

Pa *sym:* pascal.

packed column *n:* a fractionation or absorption column filled with small objects that are designed to have a relatively large surface per unit volume (the packing), instead of bubble trays or other devices, to give the required contact between the rising vapors and the descending liquid.

packed-hole assembly *n:* a drill stem that consists of stabilizers and special drill collars and is used to maintain the proper angle and course of the hole. This assembly is often necessary in crooked hole country. See *crooked hole country*.

packed pendulum assembly *n:* a bottomhole assembly in which pendulum-length collars are swung below a regular packed-hole assembly. The pendulum portion of the assembly is used to reduce hole angle; it is then removed, and the packed-hole assembly is run above the bit. See *packed-hole assembly* and *pendulum assembly*.

packer *n:* a piece of downhole equipment, consisting of a sealing device, a holding or setting device, and an inside passage for fluids, used to block the flow of fluids through the annular space between the tubing and the wall of the wellbore by sealing off the space between them. It is usually made up in the tubing string some distance above the producing zone. A sealing element expands to prevent fluid flow except through the inside bore of the packer and into the tubing. Packers are classified according to configuration, use, and method of setting and whether or not they are retrievable (that is, whether they can be removed when necessary, or whether they must be milled or drilled out and thus destroyed).

packer flowmeter *n:* a tool for production logging that employs an inflatable packer, which ensures that all the fluid from the well passes through the measuring devices built into the tool.

packer fluid *n:* a liquid, usually mud but sometimes salt water or oil, used in a well when a packer is between the tubing and casing. Packer fluid must be heavy enough to shut off the pressure of the formation being produced, must not stiffen or settle out of suspension over long periods of time, and must be noncorrosive.

packer squeeze method *n:* a squeeze cementing method in which a packer is set to form a seal between the working string (the pipe down which cement is pumped) and the casing. Another packer or a cement plug is set below the point to be squeeze-cemented. By setting packers, the squeeze point is isolated from the rest of the well. See *packer* and *squeeze cementing*.

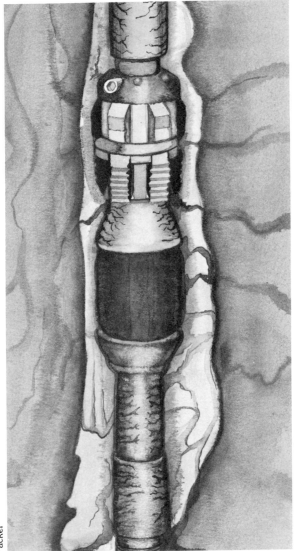

Packer

packer test *n:* a fluid-pressure test of the casing. Also called a cup test.

packing *n:* 1. a material used in a cylinder, on rotating shafts of a pump, in the stuffing box of a valve, or between flange joints to maintain a leak-proof seal. 2. the specially fabricated filling in packed fractionation columns and absorbers.

packing gland *n:* the metal part that compresses and holds packing in place in a stuffing box. See *stuffing box*.

pack off *v:* to place a packer in the wellbore and activate it so that it forms a seal between the tubing and the casing.

pad resistivity device *n:* device designed to measure the resistivity of small volumes of formation near the borehole. The instrument consists of three electrodes embedded in the center of an insulated fluid-filled rubber pad which is held against the side of the borehole. The electrodes produce two curves—one 1½ inches in depth and one 4 inches in depth. The separation between the two curves shows the difference in resistivity between the mud cake and the formation immediately behind the mud cake. The minilog is a type of pad device.

pair producton *n:* 1. the process in which a ray or wave reacts with the nucleus of an atom, converting the wave's energy into mass to produce an electron and a positron. The positron immediately reacts with an electron within the radioactive material. In the interaction, two new gamma rays are produced with energy levels less than half the original gamma ray. 2. the conversion of a photon into an electron and a positron when the photon traverses a strong electric field, such as that surrounding a nucleus or an electron.

P&A *abbr:* plug and abandon.

paraffin *n:* a hydrocarbon having the formula C_nH_{2n+2} (e.g., methane, CH_4; ethane, C_2H_6). Heavier paraffin hydrocarbons (i.e., $C_{18}H_{38}$ and heavier) form a waxlike substance that is called paraffin. These heavier paraffins often accumulate on the walls of tubing and other production equipment, restricting or stopping the flow of desirable lighter paraffins.

paraffin-base oil *n:* a crude oil characterized by a high API gravity, a high yield of low-octane gasoline, and a high yield of lubricating oil with a high pour point and a high viscosity index. Popularly, and according to an early classification system, a paraffin-base oil is a crude oil containing little or no asphalt and yielding a residue from distillation that contains paraffin wax. Compare *naphthene-base oil*.

paraffin-deposition interval *n:* an interval in the production tubing string where heavy paraffin hydrocarbons are deposited on the inside walls of the tubing. The interval is dependent on temperature; below or above certain temperatures, paraffin will not form.

paraffin hydrocarbon *n:* See *paraffin*.

paraffin inhibitor *n:* a chemical that, when injected into the wellbore, prevents or minimizes paraffin deposition.

paraffin scraper *n:* any tool used to remove paraffin from inside tubular goods.

paraffin wax *n:* a solid substance resembling beeswax but composed entirely of hydrocarbons. It is obtained from the crude wax that results from the solvent-dewaxing or cold-pressing of light paraffin distillates. The refined product is of relatively large crystalline structure, is white and brittle, and has little taste or odor.

parted rods *n pl:* sucker rods that have been broken and separated in a pumping well because of corrosion, improper loading, damaged rods, and so forth.

partial pressure *n:* the pressure exerted by one specific component of a gaseous mixture.

pascal *n:* the accepted metric unit of measurement for pressure and stress and a component in the measurement of viscosity. A pascal is equal to a force of 1 newton acting on an area of 1 square metre. Its symbol is Pa.

passivation *n:* the process of rendering a metal surface chemically inactive, either by electrochemical polarization or by contact with passivating agents.

patch *n:* a material used to cover, fill up, or mend a hole or weak spot. A metal piece extending halfway around a pipe and welded to it is a half-sole patch. Two half-sole patches make a full-sole patch.

pawl *n:* a pivoted tongue or sliding bolt on one part of a machine, adapted to fall into notches on another part (such as a ratchet wheel) to permit motion in only one direction.

pay *n:* See *pay sand*.

pay sand *n:* the producing formation, often one that is not even sandstone. It is also called pay, pay zone, and producing zone.

pay string *n:* production casing.

pay zone *n:* See *pay sand.*

PB *abbr:* plugged back; used in drilling reports.

pcf *abbr:* pounds per cubic foot.

PDC log *abbr:* perforating depth control log.

PDI *abbr:* paraffin deposition interval.

pelican hook *n:* (nautical) a wire rope attached to an anchor and sometimes to the anchor chain and used to pull and lower the anchor. The ends of the pendant not on the anchor are attached to buoys on the surface of the water.

pendulum assembly *n:* a bottomhole assembly composed of a bit and several large-diameter drill collars; it may have one or more stabilizers installed in the drill collar string. The assembly works on the principle of the pendulum effect. See *pendulum effect.*

pendulum effect *n:* the tendency of the drill stem – bit, drill collars, drill pipe, and kelly – to hang in a vertical position due to the force of gravity.

penetration rate *n:* See *rate of penetration.*

pentane *n:* any of three isomeric hydrocarbons, C_5H_{12}, of the methane series, occurring in petroleum.

pentane-plus *n:* a hydrocarbon mixture consisting mostly of normal pentane (C_5H_{12}) and heavier components, extracted from natural gas.

per *abbr:* permeability; used in drilling reports.

percussion drilling *n:* 1. cable-tool drilling. 2. rotary drilling in which a special tool called a hammer drill is used in combination with a roller cone bit.

percussion drilling tool *n:* See *hammer drill.*

perf *abbr:* perforated; used in drilling reports.

perforate *v:* to pierce the casing wall and cement to provide holes through which formation fluids may enter or to provide holes in the casing so that materials may be introduced into the annulus between the casing and the wall of the borehole. Perforating is accomplished by lowering into the well a perforating gun, or perforator, that fires electrically detonated bullets or shaped charges from the surface.

perforated completion *n:* a well completion in which the production casing or liner is punctured to allow passage between the wellbore and the producing formation. Perforations are usually made with bullet- or jet-perforating guns.

perforated liner *n:* a liner that has had holes shot in it by a perforating gun. See *liner.*

perforated pipe *n:* sections of pipe (such as casing liner, tail pipe, etc.) in which holes or slots have been cut before it is set.

perforate underbalanced *v:* to perforate the well with a column of fluid in the wellbore which exerts less pressure on bottom than the formation does, to cause formation fluids to flow into the wellbore immediately after the casing is perforated. The method is also called reverse-pressure perforating.

perforating depth control log *n:* a special type of radioactivity log that measures the depth of each casing collar. Knowing the depth of the collars makes it easy to determine the exact depth of the formation to be perforated by correlating casing-collar depth with formation depth.

perforating gun *n:* a device, fitted with shaped charges or bullets, that is lowered to the desired depth in a well and fired to create penetrating holes in casing, cement, and formation.

perforation *n:* a hole made in the casing, cement, and formation, through which formation fluids enter a wellbore. Usually several perforations are made at a time.

period of pitch *n:* the time required for the bow or the stern of a floating offshore drilling rig to start at its lowest position, rise with a wave, and return to its lowest position.

period of roll *n:* the time required for a floating offshore drilling rig to roll from one side to the other and back.

perlite *n:* a volcanic rock that can be extended to many times its original volume by crushing and heating under pressure. Release of the pressure causes expansion when the water in the rock turns to steam.

permanent completion *n:* a well completion in which production, workover, and recompletion operations can be performed without removing the wellhead.

permanent guide base *n:* a structure attached to and installed with the foundation pile when a well is drilled from an offshore drilling rig. It is seated in the temporary guide base and serves as a wellhead housing. Also, guidelines are attached to it so that equipment (such as the blowout preventers) may be guided into place on the wellhead.

permeability *n:* 1. a measure of the ease with which fluids can flow through a porous rock. 2. the fluid conductivity of a porous medium. 3. the ability of a fluid to flow within the interconnected pore network of a porous medium. See *absolute permeability*, *effective permeability*, and *relative permeability*.

persistence *n*: the durability or longevity of inhibitors used in corrosion control.

personnel net *n*: a net attached to a floatable ring, on which personnel ride when being transferred from boat to rig on offshore locations. It is usually rigged to a crane.

petrochemical *n*: a chemical manufactured from petroleum and natural gas or from raw materials derived from petroleum and natural gas.

petrol *n*: (British) gasoline.

petroleum *n*: a substance occurring naturally in the earth and composed mainly of mixtures of chemical compounds of carbon and hydrogen, with or without other nonmetallic elements such as sulfur, oxygen, and nitrogen. The compounds that compose it may be in the gaseous, liquid, or solid state, depending on their nature and on the existent conditions of temperature and pressure.

petroleum geology *n*: the study of oil- and gas-bearing rock formations. It deals with the origin, occurrence, movement, and accumulation of hydrocarbon fuels.

Petroleum Industry Training Board *n*: legislative arm of the United Kingdom government that monitors drilling training in the United Kingdom sector of the North Sea. Based at Kingfisher House, Walton Street, Aylesbury, Buckinghamshire.

Petroleum Industry Training Service (Canada) *n*: an industry-controlled and industry-operated training organization maintained specifically to assist Canadian companies with their training; headquartered at 10330-71 Avenue, Edmonton, Alberta, T6E 0W8.

petroliferous *adj*: containing petroleum (said of rocks).

phase *n*: 1. any portion of a nonhomogeneous system that is bounded by a surface, is homogeneous throughout, and may be mechanically separated from the other phases. The three phases of H_2O, for example, are ice (solid), water (liquid), and steam (gas). 2. in physics, the stage or point in a cycle to which a rotation, oscillation, or variation has advanced.

phenolics *n*: thermosetting plastic materials formed by the condensation of phenols (containing C_6H_5OH) with aldehydes (containing CHO) and used as protective coatings for oil field structures.

photoelectric effect *n*: the absorption of gamma rays, resulting in ejection of electrons from an atom. Photoelectric effect occurs when light of sufficient energy falls upon an atom and causes it to lose electrons. The energy of the light actually tears electrons away from the atoms of a substance.

photon *n*: a quantum, or unit, of electromagnetic radiation energy.

pH value *n*: a unit of measure of the acid or alkaline condition of a substance. A neutral solution (such as pure water) has a pH of 7; acid solutions are less than 7; basic, or alkaline, solutions are more than 7. The pH scale is a logarithmic scale; a substance with a pH of 4 is more than twice as acid as a substance with a pH of 5. Similarly, a substance with a pH of 9 is more than twice as alkaline as a substance with a pH of 8.

PI *abbr*: productivity index.

pickle *n*: a cylindrical or spherical device that is affixed to the end of a wireline just above the hook to keep the line straight and provide weight.

pickup position *n*: the point in drilling at which the floor crew can latch the elevators around the pipe for coming out of the hole.

piercement dome *n*: a mass of material, usually salt, that rises and penetrates rock formations.

pig *n*: 1. a scraping tool that is forced through a pipeline or flow line to clean out accumulations of wax, scale, and debris from the walls of the pipe. It travels with the flow of product in the line, cleaning the pipe walls by means of blades or brushes affixed to it. Also called a line scraper or a go-devil. 2. a batching device used to separate different products traveling in the same pipeline. It is a cylinder with neoprene or plastic cups on either end. 3. a displacement device used to displace liquid hydrocarbons from natural gas pipelines. It is a neoprene spheroid, automatically launched and received. *v*: to force a device called a pig through a pipeline or a flow line for the purpose of cleaning the interior walls of the pipe, separating different products, or displacing fluids.

piggyback *v*: (nautical) to install anchors behind each other in tandem on the same mooring line.

pig iron *n*: (slang) a piece of oil field equipment made of iron or steel.

pilot bit *n*: a bit placed on a special device called a hole opener that serves to guide the device into an already existing hole that is to be opened (made

pilot mill-pipeline oil

larger in diameter). The pilot bit merely guides, or pilots, the cutters on the hole opener into the existing hole so that the hole-opening cutters can enlarge the hole to the desired size.

pilot mill n: a special mill that has a heavy tubular extension below it called a pilot or stinger. The pilot, smaller in diameter than the mill, is designed to go inside drill pipe or tubing that is lost in the hole. It guides the mill to the top of the pipe and centers it, thus preventing the mill from bypassing the pipe.

pilot pin n: the machined extension on the very end of the bearing pin that fits into the nose of the cone of a roller cone bit.

pin n: 1. the male section of a tool joint. 2. on a bit, the bit shank. 3. one of the pegs that are fitted on each side into the link plates (side bars) of a chain link of roller chain and that serve as the stable members onto which bushings are press-fitted and around which rollers move.

pin angle n: also called journal angle. See *journal angle*.

pinch bar n: a steel lever having a pointed projection at one end and used to move a heavy load.

pinch in v: to decrease the size of the opening of an adjustable choke when a kick is being circulated out of a well.

pinch-out n: a geological structure that forms a trap for oil and gas when a porous and permeable rock ends at or stops against an impervious formation.

pin-drive master bushing n: a master bushing that has four drive holes corresponding to the four pins on the bottom of the pin-drive kelly bushing.

pinion n: 1. a gear with a small number of teeth designed to mesh with a larger wheel or rack. 2. the smaller of a pair or the smallest of a train of gear wheels.

pin link n: a link of roller chain consisting of four parts—two side bars and two pins. The pins are press-fitted into the side bars (pin link plates).

pin packer n: a packer in which the packing element is held in position by brass or steel pins. When weight is put on the packer, two metal sleeves telescope, shearing the pins and allowing the element to fold and pack off.

pin tap n: a short, threaded device made up on the bottom of drill pipe or tubing and used to screw into the box of a stand of drill pipe or drill collars lost in the hole. Once the pin tap is engaged, the lost pipe can be retrieved.

pipe n: a long hollow cylinder, usually steel, through which fluids are conducted. Oil field tubular goods are casing (including liners), drill pipe, tubing, or line pipe. Casing, tubing, and drill pipe are designated by external diameter. Because lengths of pipe are joined by external-diameter couplings threaded by standard tools, an increase in the wall thickness can be obtained only by decreasing the internal diameter. Thus, the external diameter is the same for all weights of the same-size pipe. Weight is expressed in pounds per foot or kilograms per metre. Grading depends on the yield strength of the steel.

pipe dolly n: any device equipped with rollers and used to move drill pipe or collars. It is usually placed under one end of the pipe while the pipe is being lifted from the other end by the catline.

pipe fitting n: an auxiliary part (such as a coupling, elbow, tee, cross, etc.) used for connecting lengths of pipe.

pipe hanger n: 1. a circular device with a frictional gripping arrangement used to suspend casing and tubing in a well. 2. a device used to support a pipeline.

pipe jack n: a hand tool used to lift and move a stand of pipe that is set back in the derrick. It has a handle on one end and two semicircular pieces on the other end that are designed to fit under the shoulder of a joint of pipe and avoid damage as the pipe is lifted with the tool.

pipeline n: a system of connected lengths of pipe, usually buried in the earth or laid on the seafloor, that is used for transporting petroleum and natural gas.

pipeline connection n: the outlet from a well or a tank by which oil or gas is transferred to a pipeline for transportation away from the field.

Pipe Line Contractors Association n: a national trade organization, founded in 1948 and based in Dallas, Texas, for American pipeline contractors.

pipeline gas n: gas that meets the minimum specifications of a transmission company.

pipeline gauger n: an employee of a pipeline company who measures the quantity and quality of crude oil in a tank before the oil is pumped into a pipeline.

pipeline oil n: a crude oil whose BS&W content is low enough to make the oil acceptable for pipeline shipment.

pipeline patrol *n:* a watch, usually maintained from an airplane, to check the route of a pipeline for leaks or other abnormal conditions.

pipe protector *n:* a protector that prevents drill pipe from rubbing against the hole or against the casing.

pipe rack *n:* a horizontal support for tubular goods.

pipe racker *n:* 1. (obsolete) a worker who places pipe to one side in the derrick. 2. a pneumatic or hydraulic device often used on drill ships that, upon command from an operator, either picks up pipe from a rack and lifts it into the derrick or takes pipe from out of the derrick and places it on the rack. It eliminates the need to stand pipe in the derrick while it is out of the hole.

pipe-racking fingers *n:* extensions within a pipe rack for keeping individual pipes separated.

pipe ram *n:* a sealing component for a blowout preventer that closes the annular space between the pipe and the blowout preventer or wellhead.

pipe ram preventer *n:* a blowout preventer that uses pipe rams as the closing elements. See *pipe ram*.

pipe repair clamp *n:* a clamp used to make a temporary repair of a leak in a pipeline.

pipe saddle *n:* a fitting made in parts to clamp onto a pipe to stop a leak or provide an outlet.

pipe tongs *n:* also called tongs. See *tongs*.

pipe upset *n:* that part of the pipe that has an abrupt increase of dimension.

pipe wiper *n:* a flexible disk-shaped device, usually made of rubber, with a hole in the center through which drill pipe or tubing passes, used to wipe off mud, oil, or other liquid from the pipe as the pipe is pulled from the hole.

piston *n:* a cylindrical sliding piece that is moved by or moves against fluid pressure within a confining cylindrical vessel.

piston crown *n:* See *crown*.

piston pin *n:* a pin that forms a flexible link between the piston and the connecting rod. Also called a wrist pin. This bearing area has the highest load per square inch (square millimetre) of any in an engine. Piston pin bearing loads may be as high as 50,000 psi (345 MPa).

piston ring *n:* a yielding ring, usually metal, that surrounds a piston and maintains a tight fit inside a cylinder.

piston rod *n:* 1. a metal shaft that joins the piston to the crankshaft in an engine. 2. a metal shaft in a mud pump, one end of which is connected to the piston and the other to the pony rod.

piston stroke *n:* the length of movement, in inches (millimetres), of the piston of an engine from TDC to BDC.

P.I.T.B. *abbr:* Petroleum Industry Training Board.

pitch *n:* 1. in wireline spooling, the degree of slope that the wireline travels in going from one wrap to the next. 2. in roller chain, the distance (in inches or millimetres) between the centers of two members next to one another — the distance between the centers of the bushings or rollers.

pit level *n:* height of drilling mud in the mud pits.

pit-level indicator *n:* one of a series of devices that continuously monitor the level of the drilling mud in the mud tanks. The indicator usually consists of float devices in the mud tanks that sense the mud level and transmit data to a recording and alarm device (a pit-volume recorder) mounted near the driller's position on the rig floor. If the mud level drops too low or rises too high, the alarm sounds to warn the driller that he may be either losing circulation or taking a kick.

pit-level recorder *n:* See *pit-level indicator*.

pitman *n:* the arm that connects the crank to the walking beam on a pumping unit by means of which rotary motion is converted to reciprocating motion.

Pitot tube *n:* an open-ended tube arranged to face against the current of a stream of fluid; used in measuring the velocity head of a flowing medium.

Pitot-tube meter *n:* a meter that uses a Pitot tube and a manometer or other differential-pressure mechanism to measure flowing fluids. The difference between the pressure on the Pitot tube and the static pressure is the velocity head of the flow, which is directly related to the rate of flow.

P.I.T.S. *abbr:* Petroleum Industry Training Service (Canada).

pit-volume recorder *n:* the gauge at the driller's position that records data from the pit-level indicator.

Pit Volume Totalizer *n:* trade name for a type of pit-level indicator. See *pit-level indicator*.

pk *abbr:* pink; used in drilling reports.

pkr *abbr:* packer; used in drilling reports.

pl *abbr:* pipeline; used in drilling reports.

planimeter n: an instrument for measuring the area of a plane figure. As the point on a tracing arm is passed along the outline of a figure, a graduated wheel and disk indicate the area encompassed.

plasticity n: the ability of a substance to be deformed without rupturing.

plastic squeezing n: the procedure by which a quantity of resinous material is squeezed into a sandy formation to consolidate the sand and prevent its flowing into the well. The resinous material is hardened by the addition of special chemicals, which creates a porous mass that permits oil to flow into the well but holds back the sand at the same time. See *sand consolidation*.

plastic viscosity n: an absolute flow property indicating the flow resistance of certain types of fluids. Plastic viscosity is a measure of shearing stress.

platform n: an immobile offshore structure constructed on pilings from which wells are drilled, produced, or both.

platform jacket n: a support that is firmly secured to the ocean floor and to which the legs of a platform are anchored.

play n: 1. the extent of a petroleum-bearing formation. 2. the activities associated with petroleum development in an area.

PLCA abbr: Pipe Line Contractors Association.

Plimsoll mark n: a mark placed on the side of a floating offshore drilling rig or ship denoting the maximum depth to which it may be loaded or ballasted. The line is set in accordance with local and international rules for safety of life at sea.

p-low n: a notation of the amount of pressure generated on the drill stem when the mud pumps are run at a speed slower than the speed used when drilling ahead. A p-low or several p-lows are established for use when a kick is being circulated out of the wellbore.

plug n: any object or device that blocks a hole or passageway (as a cement plug in a borehole).

plug and abandon v: to place a cement plug into a dry hole and abandon it.

plug back v: to place cement in or near the bottom of a well to exclude bottom water, to sidetrack, or to produce from a formation already drilled through. Plugging back can also be accomplished with a mechanical plug set by wireline, tubing, or drill pipe.

plug-back cementing n: a secondary-cementing operation in which a plug of cement is positioned at a specific point in the well and allowed to set.

plug container n: See *cementing head*.

plug flow n: a fluid moving as a unit in which all shear stress occurs at the pipe wall.

plugging material n: a substance used to temporarily or permanently block off zones while treating or working on other portions of the well.

plug valve n: See *valve*.

plunger n: 1. a basic component of the sucker rod pump. See *sucker rod pump*. 2. the rod that serves as a piston in a reciprocating pump.

pneumatic adj: operated by air pressure.

pneumatic control n: a control valve that is actuated by air. Several pneumatic controls are used on drilling rigs to actuate rig components (clutches, hoists, engines, pumps, etc.).

pneumatic line n: any hose or line, usually reinforced with steel, that conducts air from an air source (such as a compressor) to a component that is actuated by air (such as a clutch).

pneumofathometer n: a depth-indicating instrument used by a diver.

pod n: See *hydraulic control pod*.

Pod (Podbielniak) analysis n: an analytical procedure for hydrocarbon gases and liquids whereby the various components are quantitatively separated by low-temperature distillation for identification and measurement.

point-reaction force n: a force that counteracts another force at a single point.

poise n: the viscosity of a liquid in which a force of 1 dyne (a unit of measurement of small amounts of force) exerted tangentially on a surface of 1 cm² of either of two parallel planes 1 cm apart will move one plane at the rate of 1 cm per second in reference to the other plane, with the space between the two planes filled with the liquid.

polar compound n: a compound (such as water) with a molecule that behaves as a small bar magnet with a positive charge on one end and a negative charge on the other.

pole mast n: a portable mast constructed of tubular members. A pole mast may be a single pole, usually of two different sizes of pipe telescoped together to be moved or extended and locked to obtain maximum height above a well. Double-pole masts give added strength and stability. See *mast*.

polished rod n: the topmost portion of a string of sucker rods, used for lifting fluid by the rod-pumping method. It has a uniform diameter and is smoothly polished to effectively seal pressure in the stuffing box attached to the top of the well.

polyester *n:* a thermosetting or thermoplastic material formed by esterification of polybasic organic acids with polyhydric acids.

polymer *n:* a substance that consists of large molecules formed from smaller molecules in repeating structural units. In petroleum refining, heat and pressure are used to polymerize light hydrocarbons into larger molecules, such as those that make up high-octane gasoline. In oil field operations, various types of organic polymers are used to thicken drilling mud, fracturing fluid, acid, and other liquids. In petrochemical production, polymer hydrocarbons are used as the basis for plastics.

polymerization *n:* the bonding of two or more simple molecules to form larger molecular units.

polymer mud *n:* a drilling mud to which has been added a polymer, a chemical that consists of large molecules that were formed from small molecules in repeating structural units, to increase the viscosity of the mud.

polymer units *n pl:* units in a refinery that polymerize propylene and ethylene in the presence of a catalyst and in a liquid solvent. Polypropylene and polyethylene are produced. Following polymerization, the product is removed from the reactor and processed through a purification facility for removal of catalyst residue and separation of the solvent. The powdered product is then conveyed in a pneumatic conveyor to the finishing area where additives are blended with powder and the mixture is processed through an extruder to form pellets.

pony rod *n:* 1. a sucker rod less than 25 feet long. 2. the rod joined to the connecting rod and piston rod in a mud pump.

pool *n:* a reservoir or group of reservoirs. The term is a misnomer in that hydrocarbons seldom exist in pools but rather in the pores of rock.

poor boy *v:* to make do; to do something on a shoestring. *adj:* homemade.

POP *abbr:* putting on the pump; used in drilling reports.

popcorn *adj:* substandard, unsafe, or cheap.

poppet valve *n:* a device that controls the rate of flow of fluid in a line or opens or shuts off the flow of fluid completely. When open, the sealing surface of the valve is moved away from a seat; when closed, the sealing surface contacts the seat to shut off flow. Usually, the direction of movement of the valve is perpendicular to the seat. Poppet valves are used extensively as pneumatic (air) controls on drilling rigs and as intake and exhaust valves in most internal-combustion engines.

pop valve *n:* a spring-loaded safety valve that opens automatically when pressure exceeds the limits for which the valve is set. It is used as a safety device on pressurized vessels and other equipment to prevent damage from excessive pressure. It also is called a relief valve or a safety valve.

por *abbr:* porosity or pores; used in drilling reports.

pore *n:* an opening or space within a rock or mass of rocks, usually small and often filled with some fluid (water, oil, gas, or all three). Compare *vug*.

porosity *n:* the condition of something that contains pores (such as a rock formation). See *pore*.

port *n:* 1. (nautical) the left side of a vessel (determined by looking toward the bow). 2. the opening in the side of a liner in a two-stroke cycle engine.

portable mast *n:* a mast mounted on a truck and capable of being erected as a single unit. See *telescoping derrick*.

portland cement *n:* the cement most widely used in oilwells. It is made from raw materials such as limestone, clay or shale, and iron ore.

position-reference system *n:* any system or method by which surveillance is maintained on the position of a floating offshore drilling rig in relation to the subsea wellhead. Ideally, the rigs should always be directly over the wellhead to minimize wear on subsea equipment and facilitate operations involved with the equipment. See *acoustic position reference* and *tautline position-reference system*.

positive choke *n:* a choke in which the orifice size must be changed to change the rate of flow through the choke.

positive clutch *n:* a clutch in which jaws or claws interlock when pushed together. Two types of positive clutches are the jaw clutch and the spline clutch.

positive-displacement meter *n:* a mechanical, fluid-measuring device that measures by filling and emptying chambers of a specific volume; also known as a volume meter or volumeter. The displacement of a fixed volume of fluid may be accomplished by the action of reciprocating or oscillating pistons, rotating vanes or buckets, nutating disks, or tanks or other vessels that automatically fill and empty.

positive-displacement motor *n:* usually called a Dyna-Drill. See *Dyna-Drill*.

positive-displacement pump *n:* a pump that moves a measured quantity of liquid with each stroke of a piston or each revolution of vanes or gears; a reciprocating pump or a rotary pump.

positron *n:* a particle similar to an electron but carrying a positive charge of the same mass and magnitude as the charge of an electron. Positrons are emitted when there is an excess of protons in the nucleus of an atom.

possum belly *n:* 1. a receiving tank situated at the end of the mud return line. The flow of mud comes into the bottom of the device and travels over baffles to control mud flow over the shale shaker. 2. a metal box under a truck bed that holds pipeline repair tools.

posthole digger *n:* (slang) a small or makeshift drilling rig.

posthole well *n:* (slang) a relatively shallow well. Sometimes, a dry hole.

potential *n:* the maximum volume of oil or gas that a well is capable of producing.

potential test *n:* a test of the rate at which a well can produce oil or gas. See *potential*.

potentiometer *n:* 1. an instrument to measure electromotive forces by comparison with a known potential difference. 2. a resistor used chiefly as a voltage divider.

pounds per cubic foot *n:* a measure of the density of a substance (such as drilling fluid).

pounds per gallon *n:* a measure of the density of a fluid (such as drilling mud).

pounds per square inch gauge *n:* the pressure in a vessel or container as registered on a gauge attached to the container. This pressure reading does not include the pressure of the atmosphere outside the container.

pounds per square inch per foot *n:* a measure of the amount of pressure, in pounds per square inch, that a column of fluid (such as drilling mud) exerts upon the bottom of the column for every foot of its length. For example, 10-ppg mud exerts 0.52 psi/ft, so a column of 10-ppg mud that is 1,000 feet long exerts 520 psi at the bottom of the column. See *pressure gradient*.

pour point *n:* the lowest temperature at which a fuel will flow. For oil, the pour point is a temperature 5°F (−15°C) above that temperature at which the oil is solid.

power-driven mud pump *n:* a reciprocating pump for circulating drilling fluids, operated through cranks and connecting rods by power supplied to its crankshaft from an electric motor or internal-combustion engine. It is usually a duplex (with two cylinders) but may be triplex (with three cylinders). Most mud pumps have double-acting pistons, but some have single-acting pistons that function as plungers.

power rig *n:* also called mechanical rig. See *mechanical rig*.

power rod tongs *n:* tongs that are actuated by air or hydraulic fluid and are used for making up or breaking out sucker rods.

power-shift transmission *n:* a device of clutches and gears used with an engine for automatically and smoothly changing power ratios. It is usually used in conjunction with a hydraulic torque converter.

power slips *n:* See *slips*.

power sub *n:* a hydraulically powered device used to turn the drill pipe, tubing, or casing in a well in lieu of a rotary.

power takeoff *n:* a gearbox or other device serving to relay the power of a prime mover to auxiliary equipment.

power-tight coupling *n:* coupling screwed on casing tightly enough to be leakproof at the time of makeup.

power tongs *n:* a wrench that is used to make up or break out drill pipe, tubing, or casing on which the torque is provided by air or fluid pressure. Conventional tongs are operated by mechanical pull provided by a jerk line connected to a cathead.

pozzolan *n:* a natural or artificial siliceous material commonly added to portland cement mixtures to impart certain desirable properties. Added to oilwell cements, pozzolans reduce slurry weight and viscosity, increase resistance to sulfate attack, and influence factors such as pumping time, ultimate strength, and watertightness.

pozzolan-cement mixture *n:* a mixture of pozzolan and cement.

pozzolan-lime reaction *n:* the reaction between pozzolan and lime in the presence of water, forming a cementitious material primarily composed of hydrated calcium silicates.

ppg *abbr:* pounds per gallon.

ppm *abbr:* parts per million.

prairie dog plant *n:* a small, comparatively simple refinery located in a remote area.

prefab *n:* a windbreak used around the rig floor, engines, substructure, and other areas to protect the crew from cold winds during winter operations. Windbreaks are constructed of either canvas (tarp), wood, or metal.

preflush *n:* the quantity of fluid used ahead of the acid solution pumped into a well in an acid-stimulation treatment; sometimes called spearhead. Compare *overflush*.

preforming *n:* a process in the manufacture of wire rope that crimps the strands, giving the rope a slightly permanent set and controlling its flexibility.

preignition *n:* a condition in an internal-combustion engine characterized by a knocking sound and caused by the fuel-air mixture having been ignited too soon because of an abnormal condition.

press-fit tolerance *n:* the amount of tolerance allowed (either over or under a specific measurement) when a device, such as a bearing, is pressed into a machined receptacle.

pressure *n:* the force that a fluid (liquid or gas) exerts uniformly in all directions within a vessel, pipe, hole in the ground, and so forth, such as that exerted against the liner wall of a tank or that exerted on the bottom of the wellbore by drilling mud. Pressure is expressed in terms of force exerted per unit of area, as pounds per square inch (psi) or grams or kilograms per square centimetre.

pressure base *n:* See *base pressure*.

pressure compensator *n:* a device installed on the leg of a roller cone bit having sealed and lubricated bearings. Its function is to maintain equal pressure inside and outside the bit's bearings, in spite of the fact that the drilling mud in the hole can exert very high pressure outside the bit.

pressure differential *n:* also called differential pressure. See *differential pressure*.

pressure drop *n:* a loss of pressure, resulting from friction, sustained by a fluid passing through a line, valve, fitting, or other device.

pressure gauge *n:* an instrument that measures fluid pressure and usually registers the difference between atmospheric pressure and the pressure of the fluid by indicating the effect of such pressures on a measuring element (a column of liquid, a Bourdon tube, a weighted piston, a diaphragm, or other pressure-sensitive device).

pressure gradient *n:* a scale of pressure differences in which there is a uniform variation of pressure from point to point. For example, the pressure gradient of a column of water is about 0.433 psi/ft of vertical elevation. The normal pressure gradient in a formation is equivalent to the pressure exerted at any given depth by a column of 10 percent salt water extending from that depth to the surface (0.465 psi/ft or 10.518 kPa/m).

pressure loss *n:* 1. a reduction in the amount of force a fluid exerts against a surface, usually occurring because the fluid is moving against the surface. 2. the amount of pressure indicated by a drill pipe pressure gauge when drilling fluid is being circulated by the mud pump. Pressure losses occur as the fluid is circulated.

pressure maintenance *n:* the use of waterflooding or natural gas recycling during primary recovery to provide additional formation pressure and displacement energy that can supplement and conserve natural reservoir drives. Although commonly begun during primary production, pressure maintenance methods are often considered to be a form of enhanced oil recovery. See *enhanced oil recovery* and *secondary recovery*.

pressure parting *n:* a phenomenon in which a rock formation is broken apart along bedding planes or in which natural cracks are widened by the application of hydraulic pressure. It is sometimes called breaking or cracking the formation, earth lifting, or formation fracturing.

pressure probe *n:* a diagnostic tool used to ascertain whether there is a gas leak in the tubing of a gas-lift well. If there is a tubing leak, the pressure on the annulus will equal the pressure on the tubing.

pressure relief valve *n:* a valve that opens at a preset pressure to relieve excessive pressures within a vessel or line. Also called a relief valve, safety valve, or pop valve.

pressure storage tank *n:* a storage tank constructed to withstand pressure generated by the vapors inside. Such a tank is often spherical, has a wall thickness greater than that of the usual storage tank, and has a concave or a convex top.

pressure vessel *n:* any container designed to contain fluids at a pressure substantially greater than atmospheric.

pressure, volume, and temperature analysis *n:* an examination of reservoir fluid in a laboratory under various pressures, volumes, and temperatures to determine the characteristics and behavior of the fluid.

pressurize *v:* to increase the internal pressure of a closed vessel.

preventer *n:* shortened form of blowout preventer. See *blowout preventer.*

preventive maintenance *n:* a system of conducting regular checks and testing of equipment to permit replacement or repair of weakened or faulty parts before failure of the equipment results.

primary cementing *n:* the cementing operation that takes place immediately after the casing has been run into the hole; used to provide a protective sheath around the casing, to segregate the producing formation, and to prevent the undesirable migration of fluids. See *secondary cementing* and *squeeze cementing.*

primary porosity *n:* natural porosity in petroleum reservoir sand or rocks; the porosity developed during the original sedimentation process by which the rock was formed. Formations having this property (such as sand) are usually granular.

primary recovery *n:* the first stage of oil production in which natural reservoir drives are used to recover oil. When natural drives are still operative but incapable of producing profitable amounts of oil, they may be boosted by various types of artificial lift, well stimulation, and pressure maintenance. See *artificial lift, well stimulation,* and *pressure maintenance.*

prime mover *n:* an internal-combustion engine that is the source of power for a drilling rig in oilwell drilling.

probe *n:* any small device that, when brought into contact with or inserted into a system, can make measurements on that system. In corrosion, probes can measure electrical potential or the corrosivity of various substances to determine a system's corrosive tendencies.

producing zone *n:* the zone or formation from which oil or gas is produced. See *pay sand.*

production *n:* 1. the phase of the petroleum industry that deals with bringing the well fluids to the surface and separating them and with storing, gauging, and otherwise preparing the product for the pipeline. 2. the amount of oil or gas produced in a given period.

production casing *n:* the last string of casing liner that is set in a well, inside of which is usually suspended the tubing string.

production log *n:* also called a spinner survey. See *spinner survey.*

production packer *n:* any packer designed to make a seal between the tubing and the casing during production.

production platform *n:* See *platform.*

production rig *n:* a portable servicing or workover outfit, usually mounted on wheels and self-propelled. A well-servicing unit consists of a hoist and engine mounted on a wheeled chassis with a self-erecting mast. A workover rig is basically the same, with the addition of a substructure with rotary, pump, pits, and auxiliaries to permit handling and working a drill string.

production tank *n:* a tank used in the field to receive crude oil as it comes from the well. Also called a flow tank or lease tank.

productivity index *n:* a well-test measurement indicative of the amount of oil or gas a well is capable of producing. It may be expressed as

$$PI = \frac{Q}{P_s - P_f}$$

where

PI = productivity index (b/d or Mcf/d per psi of pressure differential)
Q = rate of production (b/d or Mcf/d)
P_s = static bottomhole pressure (psi)
P_f = flowing bottomhole pressure (psi).

products cycle *n:* the sequence or order in which a number of different products are batched through a pipeline.

products line *n:* a pipeline used to ship refined products.

propane *n:* a paraffin hydrocarbon (C_3H_8) that is a gas at ordinary atmospheric conditions but is easily liquefied under pressure. It is a constituent of liquefied petroleum gas.

propane, commercial *n:* See *commercial propane.*

proppant *n:* also called propping agent. See *propping agent.*

propping agent *n:* a granular substance (sand grains, aluminum pellets, or other material) that is carried in suspension by the fracturing fluid and that serves to keep the cracks open when fracturing fluid is withdrawn after a fracture treatment.

propylene *n:* the chemical compound of the olefin series having the formula C_3H_6. Its official name is propene.

proration *n:* a system, enforced by a state or federal agency or by agreement between operators, that limits the amount of petroleum that can be produced from a well or a field within a given period.

protection casing *n:* a string of casing set deeper than the surface casing to protect a section of the hole and to permit drilling to continue to a greater depth. Sometimes called intermediate casing string.

protium *n:* an isotope of hydrogen with no neutrons in the nucleus, designated as $_1H^1$. It is the lightest isotope of hydrogen, known as light hydrogen.

proton *n:* a particle in the nucleus of an atom that has a positive charge. The total positive charge of the protons in the nucleus of an atom is equal to the total negative charge of the electrons.

prove *v:* to determine the accuracy of a petroleum measurement meter.

prover *n:* a device used to determine the accuracy of a petroleum measurement meter.

prover tank *n:* a tank used to calibrate liquid flowmeters.

psi *abbr:* pounds per square inch.

psia *abbr:* pounds per square inch absolute. Psia is equal to the gauge pressure plus the pressure of the atmosphere at that point.

psi/ft *abbr:* pounds per square inch per foot.

psig *abbr:* pounds per square inch gauge.

PTO *abbr:* power takeoff.

puddling *n:* 1. in cement evaluation work, the agitation of cement slurry with a rod to remove trapped air bubbles. 2. in field practice, the rotation of the casing during or after a primary cementing operation.

pull a well *v:* to remove rods or tubing from a well.

pull a well in *v:* to collapse a derrick.

pull casing *v:* to remove casing from a well.

pull-down *n:* a snubbing unit; a device used to apply additional force to the drill stem or tubing when it is necessary to put the drill stem or tubing into the hole against high pressure.

pulley *n:* a wheel with a grooved rim, used for pulling or hoisting. See *sheave*.

pulling tool *n:* a hydraulically operated tool that is run in above the fishing tool and anchored to the casing by slips. It exerts a strong upward pull on the fish by hydraulic power derived from fluid that is pumped down the fishing string.

pulling unit *n:* a well-servicing outfit used in pulling rods and tubing from the well. See *production rig*.

pull it green *v:* to pull a bit from the hole for replacement before it is greatly worn.

pull out *v:* See *come out of the hole*.

pulsation dampener *n:* any gas- or liquid-charged, chambered device that minimizes periodic increases and decreases in pressure (as from a mud pump).

pulsed-neutron survey *n:* a special cased-hole logging method that uses radioactivity reaction time to obtain measurements of water saturation, residual oil saturation, and fluid contacts in the formation outside the casing of an oilwell.

pulse-echo techniques *n:* corrosion-detecting processes which, by recording the action of ultrasonic waves artificially introduced into production structures, can determine metal thicknesses and detect flaws.

pump *n:* a device that increases the pressure on a fluid or raises it to a higher level. Various types of pumps include the reciprocating pump, centrifugal pump, rotary pump, jet pump, sucker rod pump, hydraulic pump, mud pump, submersible pump, and bottomhole pump.

pumpability *n:* the physical characteristic of a cement slurry that determines its ability to be pumped.

pump barrel *n:* the cylinder or liner in which the plunger of a sucker rod pump reciprocates. See *sucker rod pump* and *working barrel*.

pumper *n:* the oil company employee who attends to producing wells. He supervises any number of wells, ensuring steady production, preparing reports, testing, gauging, and so forth. Also called a switcher or lease operator.

pump house *n:* a building that houses the pumps, engines, and control panels at a pipeline gathering station or trunk station.

pumping tee *n:* a heavy-duty steel, T-shaped pipe fitting that is screwed or flanged to the top of a pumping well. The polished rod works through a stuffing box on top of the tee and in the run of the tee to operate a sucker rod pump in the well. Pumped fluid is discharged through the side opening of the tee.

pumping unit *n:* the machine that imparts reciprocating motion to a string of sucker rods extending to the positive-displacement pump at the bottom of a well; usually a beam arrangement driven by a crank attached to a speed reducer.

pump jack *n:* a surface unit similar to a pumping unit but having no individual power plant. Usually several pump jacks are operated by pull rods or cables from one central power source.

pump liner n: a cylindrical, accurately machined, metallic section that forms the working barrel of some reciprocating pumps. Liners are an inexpensive means of replacing worn cylinder surfaces, and in some pumps they provide a method of conveniently changing the displacement and capacity of the pumps.

pump manifold n: an arrangement of valves and piping that permits a wide choice in the routing of suction and discharge fluids among two or more pumps.

pump off v: to pump (a well) so that the fluid level drops below the standing valve of the pump and it stops working.

pump pressure n: fluid pressure arising from the action of a pump.

pump station n: one of the installations built at intervals along an oil pipeline to contain storage tanks, pumps, and other equipment to route and maintain the flow of oil.

pump valve n: any of the valves on a reciprocating pump (such as suction and discharge valves) or on a sucker rod pump (such as a ball-and-seat valve).

pup joint n: a length of drill pipe, tubing, or casing shorter than 30 feet.

pure fatigue n: metal fatigue for which no cause can be determined.

pusher n: shortened form of toolpusher.

pushrod n: a device used to link the valve to the cam in an engine.

put a well on v: to make a well start flowing or pumping.

put on the pump v: to install a pump jack or pumping unit, sucker rods, and bottomhole pump in a well.

PVT abbr: 1. Pit Volume Totalizer. 2. pressure, volume, and temperature. See *pressure, volume, and temperature analysis*.

PVT analysis abbr: pressure, volume, and temperature analysis.

pyrometer n: an instrument for measuring temperatures, especially those above the range of mercury thermometers.

qtz *abbr:* quartz; used in drilling reports.

qtze *abbr:* quartzite; used in drilling reports.

quantum *n:* a unit of energy.

quartz *n:* a hard mineral composed of silicon dioxide; a common component in igneous, metamorphic, and sedimentary rocks.

quartzite *n:* a compact granular rock composed of quartz and derived from sandstone by metamorphism.

quebracho *n:* a South American tree that is a source of tannin extract, which has been extensively used as a thinning agent for drilling mud, but is seldom used today.

quench *v:* to cool heat-treated metal rapidly by immersion in an oil or water bath.

quicklime *n:* unslaked lime (calcium oxide).

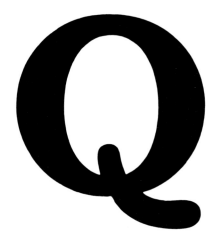

quick-setting cement *n:* a lightweight slurry designed to control lost circulation by setting very quickly.

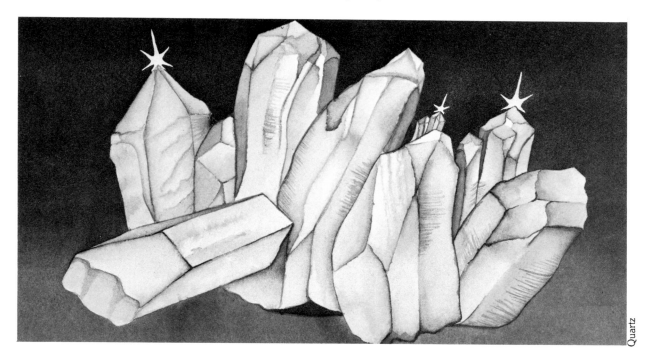

Quartz

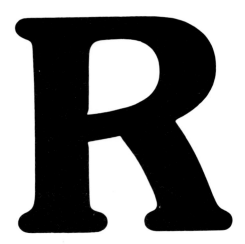

R *abbr:* Rankine. See *Rankine temperature scale.*

rabbit *n:* 1. a small plug that is run through a flow line to clean the line or to test for obstructions. 2. any plug left unintentionally in a pipeline during construction (as, a rabbit that ran into the pipe).

race *n:* a groove for the balls in a ball bearing or for the rollers in a roller bearing.

rack *n:* 1. framework for supporting or containing a number of loose objects, such as pipe. See *pipe rack*. 2. a bar with teeth on one face for gearing with a pinion or worm gear. 3. a notched bar used as a ratchet. *v:* 1. to place on a rack. 2. to use as a rack.

racking platform *n:* a small platform with fingerlike steel projections attached to the side of the mast on a well-servicing unit. When a string of sucker rods or tubing is pulled from a well, the top end of the rods or tubing is placed (racked) between the steel projections and held in a vertical position in the mast.

rack pipe *v:* 1. to place pipe withdrawn from the hole on a pipe rack. 2. to stand pipe on the derrick floor when coming out of the hole.

radial flow *n:* the flow pattern of fluids flowing into a wellbore from the surrounding drainage area.

radiation logging *n:* See *radioactivity well logging.*

radiator *n:* an arrangement of pipes that contains a circulating fluid and is used for heating an external object or cooling an internal substance by radiation.

radioactive *adj:* of, caused by, or exhibiting radioactivity.

radioactive tracer *n:* a radioactive material (often carnotite) put into a well to allow observation of fluid or gas movements by means of a tracer survey.

radioactivity *n:* the property possessed by some substances (radium, uranium, or thorium) of releasing alpha particles, beta particles, or gamma particles as the substance spontaneously disintegrates.

radioactivity log *n:* a record of the natural or induced radioactive characteristics of subsurface formations. See *radioactivity well logging.*

Rotary

radioactivity well logging *n:* the recording of the natural or induced radioactive characteristics of subsurface formations. A radioactivity log, also known as a radiation log, normally consists of two recorded curves: a gamma ray curve and a neutron curve. Both help to determine the types of rocks in the formation and the types of fluids contained in the rocks. The two logs may be run simultaneously in conjunction with a collar locator in a cased or uncased hole.

radiographic examination *n:* photographic record of corrosion damage obtained by transmitting X rays or radioactive isotopes into production structures.

ram *n:* the closing and sealing component on a blowout preventer. One of three types – blind, pipe, or shear – may be installed in several preventers mounted in a stack on top of the wellbore. Blind rams, when closed, form a seal on a hole that has no drill pipe in it; pipe rams, when closed, seal around the pipe; shear rams cut through drill pipe and then form a seal.

ram blowout preventer *n:* a blowout preventer that uses rams to seal off pressure on a hole that is with or without pipe. Also called a ram preventer.

ram preventer *n:* also called a ram blowout preventer. See *ram blowout preventer.*

range length *n:* a grouping of pipe lengths. API designation of range lengths is as follows:

	Range 1 (feet)	Range 2 (feet)	Range 3 (feet)
Casing	16-25	25-34	34 or more
Drill pipe	18-22	27-30	38-45
Tubing	20-24	28-32	

range of load *n:* in sucker rod pumping, the difference between the polished rod peak load on the upstroke and the minimum load on the downstroke.

range of stability *n:* the maximum angle to which a ship or mobile offshore drilling rig may be inclined and still be returned to its original upright position.

Rankine temperature scale *n:* a temperature scale with the degree interval of the Fahrenheit scale and the zero point at absolute zero. On the Rankine scale, water freezes at 491.60° and boils at 671.69°. See *absolute temperature scale.*

rasp *n:* a mill used in fishing operations, before running the fishing tool, to reduce the size of the box or collar on the lost tool.

rate of penetration *n:* a measure of the speed at which the bit drills into formations, usually expressed in feet (metres) per hour or minutes per foot (metre).

rathole *n:* 1. a hole in the rig floor, 30-35 feet (9-11 m) deep, which is lined with casing that projects above the floor and into which the kelly and swivel are placed when hoisting operations are in progress. 2. a hole of a diameter smaller than the main hole and drilled in the bottom of the main hole. *v:* to reduce the size of the wellbore and drill ahead.

rathole connection *n:* the addition of a length of drill pipe or tubing to the active string. The length to be added is placed in the rathole, made up to the kelly, pulled out of the rathole, and made up into the string.

raw crude *n:* a crude oil before it is refined.

raw gas *n:* unprocessed gas or the inlet gas to a plant.

raw gasoline *n:* gasoline extracted from wet natural gas.

raw mix liquids *n pl:* a mixture of natural gas liquids prior to fractionation. Also called raw make.

ray *n:* energy in wave form rather than in particle form.

RDX *n:* See *cyclonite.*

ream *v:* to enlarge the wellbore by drilling it again with a special bit. Often a rathole is reamed or opened to the same size as the main wellbore. See *rathole.*

reamer *n:* a tool used in drilling to smooth the wall of a well, enlarge the hole to the specified size, help stabilize the bit, straighten the wellbore if kinks or doglegs are encountered, and drill directionally. See *ream.*

reamer pad *n:* on a diamond bit, a flattened place above the bottomhole cutting surfaces whose purpose is to ream (enlarge) the hole above the bottom of the bit.

reboiler *n:* the auxiliary equipment to a fractionator or other column that supplies heat to the column.

rec *abbr:* recovered; used in drilling reports.

reciprocating compressor *n:* a type of compressor that has pistons moving back and forth in cylinders, suction valves, and discharge valves; a positive-displacement compressor. Reciprocating compressors are used extensively in the transmission of natural gas through pipelines.

reciprocating pump *n:* a pump consisting of a piston that moves back and forth or up and down in a cylinder. The cylinder is equipped with inlet (suction) and outlet (discharge) valves. On the intake stroke, the suction valves are opened, and fluid is drawn into the cylinder. On the discharge stroke, the suction valves close, the discharge valves open, and fluid is forced out of the cylinder.

reciprocation *n:* a back-and-forth movement (as the movement of a piston in an engine or pump).

reclaimer *n:* a system in which undesirable high-boiling contaminants of a stream are separated from the desired lighter materials; a purifying still.

recompression *n:* increasing the ambient pressure on a diver for the primary purpose of treating decompression sickness.

recompression chamber *n:* See *deck decompression chamber*.

recording gauge *n:* a device that provides a chronological record of gauge indications (as by tracing values of pressure, vacuum, voltage) on a paper form. It is driven by a clockwork mechanism.

recovery *n:* the total volume of hydrocarbons that has been or is anticipated to be produced from a well or field.

recovery factor *n:* the percentage of oil or gas in place in a reservoir that ultimately can be withdrawn by primary and/or secondary techniques.

rectifier *n:* a device used to convert alternating current into direct current.

rectifying *v:* changing an alternating current to a direct current.

red bed *n:* a layer of sedimentary rock that is predominantly red, especially one of the Permian or Triassic age.

red-lime mud *n:* a water-base clay mud containing caustic soda and tannates to which lime has been added.

reducing elbow *n:* a fitting that makes an angle between two joints of pipe and that decreases in diameter from one end to the other.

reducing flange *n:* a flange fitting used to join pipes of different diameters.

reducing nipple *n:* a pipe fitting that is threaded on both ends and decreases in diameter from one end to the other.

reducing tee *n:* a T-shaped pipe fitting with openings of two different sizes.

reduction *n:* a chemical reaction in which a compound gains electrons and obtains a more negative charge.

Redwood viscosity *n:* a unit of viscosity measurement, expressed in seconds, obtained when using a Redwood viscometer. It is the standard of viscosity measurement in Great Britain.

reef *n:* 1. a type of reservoir trap composed of rock (usually limestone) formed from the bodies of marine animals. 2. a buried coral or other reef from which hydrocarbons may be withdrawn.

reel *n:* a revolving device (such as a flanged cylinder) for winding or unwinding something flexible (such as rope or wire).

reeve *v:* to pass (as a rope) through a hole or opening in a block or similar device.

reeve the line *v:* to string a wire rope drilling line through the sheaves of the traveling and crown blocks to the hoisting drum.

reface *v:* to renew a faced surface by recutting or regrinding.

reference point *n:* also called gauge point. See *gauge point*.

refine *v:* to manufacture petroleum products from crude oil.

refinery *n:* the physical plant and attendant equipment used in the process of refining.

refinery gas *n:* the gas produced from certain petroleum refinery operations (such as cracking or reforming). The composition of refinery gas varies in accordance with the process by which it is produced, but it consists essentially of the same paraffin hydrocarbons as natural gas plus olefins (propylene, butylene, and ethylene) not found in natural gas.

refining *n:* fractional distillation of petroleum products, usually followed by other processing such as cracking.

reflux *n:* in the distillation process, that part of the condensed overhead stream that is returned to the fractionating column as a source of cooling.

reflux ratio *n:* a relative measurement of the volume of reflux in the distillation process. The ratio is commonly expressed as the quantity of reflux divided by the quantity of net overhead product.

reforming *n:* a cracking process in which low-octane naphthas or gasolines are converted into high-octane products. Thermal reforming is carried out at high temperatures and pressures (932° to 1,040°F, 250 to 1,000 psi). Catalytic reforming is carried out at lower temperatures (850° to 950°F) and much lower pressures. Reforming is usually a once-through process.

refraction *n:* deflection from a straight path undergone by a light ray or energy wave in passing from one medium to another in which the wave velocity is different, such as the bending of light rays when passing from air into water.

refracturing *n:* fracturing a formation again. See *formation fracturing.*

registered breadth *n:* the width of the hull of a mobile offshore drilling rig or a ship, measured at its greatest width and used to determine its registered tonnage.

regular cement *n:* See *common cement.*

regulator *n:* a device that reduces the pressure or volume of a fluid flowing in a line and maintains the pressure or volume at a specified level.

relative density *n:* the ratio of the mass of a given volume of a substance to the mass of a like volume of a standard substance, such as water or air. In conventional measurement units, specific gravity is similar to relative density.

relative humidity *n:* the ratio of the amount of water vapor in the air to the amount it would contain if completely saturated at a given temperature and pressure. See *absolute humidity.*

relative permeability *n:* a measure of the ability of two or more fluids (such as water, gas, and oil) to flow through a rock formation when the formation is totally filled with several fluids. The permeability measure of a rock filled with two or more fluids is different from the permeability measure of the same rock filled with only a single fluid. Compare *absolute permeability.*

release *n:* a statement filed by the lessee of an oil and gas lease indicating that the lease has been relinquished.

relief valve *n:* also called pressure relief valve. See *pressure relief valve.*

relief well *n:* a well drilled near and deflected into a well that is out of control, making it possible to bring the wild well under control. See *wild well.*

remaining on board *adj* or *n:* usually referred to by its abbreviation, ROB. Sometimes used as an adjective (cargo remaining on board), but more often used as a noun (estimating the ROB).

remote BOP control panel *n:* a device, placed on the rig floor, that can be operated by the driller to direct air pressure to actuating cylinders that turn the control valves on the main BOP control unit, located at a safe distance from the rig.

remote choke panel *n:* a set of controls, usually placed on the rig floor, that is manipulated to control the amount of drilling fluid being circulated out through the choke manifold. This procedure is necessary when a kick is being circulated out of a well. See *choke manifold.*

remote reading gauge *n:* an instrument that provides indications of pressure, vacuum, voltage, and so forth at a point distant from where the indications are actually taken.

repressure *v:* to increase or maintain reservoir pressure by injecting a pressurized fluid (such as air, gas, or water) to effect greater ultimate recovery.

reserve buoyancy *n:* the buoyancy above the waterline that keeps a floating vessel upright or seaworthy when the vessel is subjected to wind, waves, currents, and other forces of nature or when the vessel is subjected to accidental flooding.

reserve pit *n:* 1. (obsolete) a mud pit in which a supply of drilling fluid was stored. 2. a waste pit, usually an excavated earthen-walled pit. It may be lined with plastic to prevent contamination of the soil.

reserves *n pl:* the unproduced but recoverable oil or gas in a formation that has been proved by production.

reservoir *n:* a subsurface, porous, permeable rock body in which oil and/or gas is stored. Most reservoir rocks are limestones, dolomites, sandstones, or a combination of these. The three basic types of hydrocarbon reservoirs are oil, gas, and condensate. An oil reservoir generally contains three fluids – gas, oil, and water – with oil the dominant product. In the typical oil reservoir, these fluids occur in different phases because of the variance in their gravities. Gas, the lightest, occupies the upper part of the reservoir rocks; water, the lower part; and oil, the intermediate section. In addition to its occurrence as a cap or in solution, gas may accumulate independently of the oil; if so, the reservoir is called a gas reservoir. Associated with the gas, in most instances, are salt water and some oil. In a condensate reservoir, the hydrocarbons may exist as a gas, but, when brought to the surface, some of the heavier ones condense to a liquid.

reservoir drive mechanism *n:* the process in which reservoir fluids are caused to flow out of the reservoir rock and into a wellbore by natural energy. Gas drives depend on the fact that, as the reservoir is produced, pressure is reduced, allowing the gas to expand and provide the driving energy. Water-drive reservoirs depend on water pressure to force the hydrocarbons out of the reservoir and into the wellbore.

reservoir pressure n: the pressure in a reservoir.

reservoir rock n: a permeable rock that contains oil or gas in appreciable quantity.

residual fuel n: See residuals.

residuals n pl: the heavy refined hydrocarbons that are used as fuels. Bunker C oil is an example of a residual.

residue gas n: a casinghead gas that has been stripped of its gasoline.

resin cement n: an oilwell cement that is composed of resins, water, and portland cement and that provides an improved cement bond. It is mainly used in remedial operations, because its high cost prohibits its use for routine cementing of casing.

resistance n: opposition to the flow of direct current caused by a particular material or device. Resistance is equal to the voltage drop across the circuit divided by the current through the circuit.

resistivity n: the electrical resistance offered to the passage of current; the opposite of conductivity.

resistivity well logging n: the recording of the resistance of formation water to natural or induced electrical current. The mineral content of subsurface water allows it to conduct electricity. Rock, oil, and gas are poor conductors. Resistivity measurements can be correlated to formation lithology, porosity, permeability, and saturation and are very useful in formation evaluation. See electric well log.

retainer n: a cast-iron magnesium drillable tool consisting of a packing assembly and a back-pressure valve. It is used to close off the annular space between tubing or drill pipe and casing to allow the placement of cement or fluid through the tubing or drill pipe at any predetermined point behind the casing or liner, around the shoe, or into the open hole around the shoe.

retarded cement n: a cement in which the thickening time is extended by adding a chemical retarder.

retarder n: a substance added to cement to prolong the setting time so that the cement can be pumped into place. Retarders are used for cementing in high-temperature formations.

retractable bit n: a bit that can be changed by wireline operations without withdrawing the drill string. Field tests have indicated its economic feasibility, but its practicability is undetermined.

retrograde condensation n: in reservoir mechanics, the formation of liquid droplets in a gas as the well is produced and the pressure drops. Some hydrocarbons exist naturally above their critical temperature in the reservoir; as a result, when pressure is decreased, instead of expanding to form a gas, they condense to form a liquid.

return bend n: a U-shaped section of piping that connects two other pipes parallel to each other.

returns n pl: the mud, cuttings, and so forth that circulate up the hole to the surface.

reverse circulation n: the course of drilling fluid downward through the annulus and upward through the drill stem, in contrast to normal circulation in which the course is downward through the drill stem and upward through the annulus. Seldom used in open hole, but frequently used in workover operations. Also referred to as "circulating the short way", since returns from bottom can be obtained more quickly than in normal circulation.

reverse-circulation junk basket n: a special device that is lowered into the hole during normal circulation to a position over the junk to be retrieved. A ball is then pumped down to cause the drilling fluid to exit through nozzles in the tool, producing reverse circulation and creating a vacuum inside the tool so that the junk is sucked into it.

reverse emulsion n: a relatively rare oil field emulsion composed of globules of oil dispersed in water. Most oil field emulsions consist of water dispersed in oil.

reverse-pressure perforating n: See perforate underbalanced.

rheology n: the study of the flow of gases and liquids, of special importance to mud engineers and reservoir engineers.

rheostat n: a resistor that is used to vary the electrical current flow in a system.

rich amine n: the amine leaving the bottom of the contactor. It is the lean amine plus the acid gases removed from the gas by the lean amine.

rich gas n: a gas that is suitable as feed to a gas processing plant and from which products can be extracted.

rich oil n: a lean oil that has absorbed heavier hydrocarbons from natural gas.

rich-oil demethanizer n: a vessel used in gas processing plants to remove methane from rich oil.

rifle boring n: a hole bored through a machined metal piece (as in a steel pin in a traveling block sheave) through which a lubricant travels.

rig *n:* the derrick or mast, drawworks, and attendant surface equipment of a drilling or workover unit.

rig crewman *n:* also called a rotary helper. See *rotary helper*.

rig down *v:* to dismantle a drilling rig and auxiliary equipment following the completion of drilling operations; also called tear down.

rig floor *n:* the area immediately around the rotary table and extending to each corner of the derrick or mast; the area immediately above the substructure on which the drawworks, rotary table, and so forth rest. Also called derrick floor and drill floor.

right-of-way *n:* a strip of land, usually 50 to 80 feet wide, on which permission has been granted by the landowner for construction of a pipeline or road.

rig irons *n pl:* the metal parts (with the exception of nails, bolts, guy wires, and sand lines) used in the construction of a standard cable-tool rig.

rig manager *n:* an employee of a drilling contractor who is in charge of the entire drilling crew and the drilling rig. Also called a toolpusher, drilling foreman, rig supervisor, or rig superintendent.

rig superintendent *n:* also called a toolpusher. See *toolpusher*.

rig up *v:* to prepare the drilling rig for making hole; to install tools and machinery before drilling is started.

ring *n:* See *piston ring*.

ring grooves *n:* the grooves that hold the piston rings in pistons.

ring-joint flange *n:* a special type of flanged connection in which a metal ring (resting in a groove in the flange) serves as a pressure seal between the two flanges.

ringworm corrosion *n:* a form of corrosion sometimes found in the tubing of condensate wells. It occurs in a ring a few inches from the upset. Cause of ringworm corrosion has been traced to the upsetting process, in which heat required in upsetting causes the heated end to have a different grain structure from the rest of the pipe. Normalizing prevents this condition.

riser *n:* a pipe through which liquid travels upward; a riser pipe. See *riser pipe*.

riser angle indicator *n:* an acoustic or electronic device used to monitor the angle of the flex joint on a floating offshore drilling rig. Usually, a small angle should be maintained on the flex joint to minimize drill pipe fatigue and wear and damage to the blowout preventers and to maximize the ease with which tools may be run. Also called azimuth angle indicator.

riser pipe *n:* the pipe and special fittings used on floating offshore drilling rigs to establish a seal between the top of the wellbore, which is on the ocean floor, and the drilling equipment, located above the surface of the water. A riser pipe serves as a guide for the drill stem from the drilling vessel to the wellhead and as a conductor of drilling fluid from the well to the vessel. The riser consists of several sections of pipe and includes special devices to compensate for any movement of the drilling rig caused by waves. It is also called a marine riser.

riser tensioner line *n:* a cable that supports the marine riser while compensating for vessel movement.

rmg *abbr:* reaming; used in drilling reports.

ROB *abbr:* remaining on board.

rock *n:* an aggregate of different minerals. Rocks are divided into three groups on the basis of their mode of origin: igneous, metamorphic, and sedimentary.

rock a well *v:* to initiate flow by alternately bleeding pressure from, and closing off the casing and tubing of, a well that is loaded up.

rock bit *n:* also called roller cone bit. See *roller cone bit*.

rocker arm *n:* a bell-crank device that transmits the movement of the pushrod to the valves in the engine.

rock hound *n:* (slang) a geologist.

ROD *abbr:* rich-oil demethanizer.

rod *n:* See *sucker rod*.

rod back-off wheel *n:* a device used to unscrew rods when the pump is stuck or sanded up and the rods and tubing must be pulled together.

rod blowout preventer *n:* a ram device used to close the annular space around the polished rod or sucker rod in a pumping well.

rod elevators *n pl:* devices used to pull or to run sucker rods. They have a bail attached to the rod hook.

rod hanger *n:* a device used to hang sucker rods on the mast or in the derrick.

rod hook *n:* a small swivel hook having a fast-operating automatic latch to close the hook opening when weight is suspended from the hook.

rod pump *n:* See *sucker rod pump*.

rod rotor *n:* a ratchet mechanism that is actuated by a fixed rod or chain connected to the walking beam of a pumping unit and that provides a slow rate of rotation to the rod string, distributing the wear on both rods and tubing.

rod score *n:* a scratch on the surface of a sucker rod or a piston rod.

rod string *n:* a sucker rod string; the entire length of sucker rods, which usually consists of several single rods screwed together. The rod string serves as a mechanical link from the beam pumping unit on the surface to the sucker rod pump near the bottom of the well.

rod stripper *n:* a device used when rods are coated with heavy oil or when the well may flow through the tubing while the rods are being pulled. It is a form of blowout preventer.

rod sub *n:* a short length of sucker rod that is attached to the top of the sucker rod pump.

rod-transfer elevator *n:* a special type of elevator designed to accommodate the end of a sucker rod; it allows the derrickman to transfer the rod to the racking platform from the regular elevator being used to lift the rod out of the well.

rod wax *n:* a paraffin wax that forms on the sucker rod string.

rod whip *n:* the rapid, whiplike motion of the rods in a sucker rod pumping system, caused by vibration of the rod string.

rod wrench *n:* a special wrench designed for spinning up and hammering tight the joints between sucker rods. Also called a key. See *key*.

roll *n:* the angular motion of a ship or floating offshore drilling rig as its sides move up and down.

roll-dampening tanks *n pl:* the compartments of a floating offshore drilling rig that are filled with water to offset the tendency of the rig to roll.

roller bearing *n:* a bearing in which the journal rotates in contact with a number of rollers usually contained in a cage. Compare *ball bearing*.

roller chain *n:* a type of chain that is used to transmit power by fitting over sprockets attached to shafts, causing rotation of one shaft by the rotation of another. Transmission roller chain consists of roller links, pin links, and offset links.

roller cone bit *n:* a drilling bit made of two, three, or four cones, or cutters, that are mounted on extremely rugged bearings. Also called rock bits. The surface of each cone is made up of rows of steel teeth or rows of tungsten carbide inserts.

roller link *n:* one of the links in a roller chain. It consists of two bushings press-fitted into the link plates (side bars) and two rollers that fit over the bushings. The bushings are locked into the link plates to prevent rotation.

roller race *n:* a track, channel, or groove in which roller bearings roll.

ROP *abbr:* rate of penetration.

rotary *n:* the machine used to impart rotational power to the drill stem while permitting vertical movement of the pipe for rotary drilling. Modern rotary machines have a special component, the rotary bushing, to turn the kelly bushing, which permits vertical movement of the kelly while the stem is turning.

rotary bushing *n:* also called master bushing. See *master bushing*.

rotary drilling *n:* a drilling method in which a hole is drilled by a rotating bit to which a downward force is applied. The bit is fastened to and rotated by the drill stem, which also provides a passageway through which the drilling fluid is circulated. Additional joints of drill pipe are added as drilling progresses.

rotary helper *n:* a worker on a drilling or workover rig, subordinate to the driller, whose primary work station is on the rig floor. On rotary drilling rigs, there are at least two and usually three or more rotary helpers on each crew. Sometimes called floorman, roughneck, or rig crewman.

rotary hose *n:* a reinforced flexible tube on a rotary drilling rig that conducts the drilling fluid from the mud pump and standpipe to the swivel and kelly; also called the mud hose or the kelly hose.

rotary line *n:* also called drilling line. See *drilling line*.

rotary pump *n:* a pump that moves fluid by positive displacement, using a system of rotating vanes, gears, or lobes. The vaned pump has vanes extending radially from a rotating element mounted in the casing. The geared rotary pump uses oppositely rotating, meshing gears or lobes.

rotary shoe *n:* a length of pipe whose bottom edge is serrated or dressed with a hard cutting material and which is run into the wellbore around the outside of stuck casing, pipe, or tubing to mill away the obstruction.

rotary-shouldered connection *n:* the threaded and shouldered joint used in rotary drilling to join the various components of the drill stem.

rotary slips *n pl:* also called slips. See *slips*.

rotary table *n:* the principal component of a rotary, or rotary machine, used to turn the drill stem and support the drilling assembly. It has a beveled gear arrangement to create the rotational motion and an opening into which bushings are fitted to drive and support the drilling assembly.

rotary tongs *n pl:* also called tongs. See *tongs.*

rotate on bottom *v:* also called "make hole". See *make hole.*

rotating blowout preventer *n:* also called rotating head. See *rotating head.*

rotating components *n:* those parts of the drilling or workover rig that are designed to turn or rotate the drill stem and bit—swivel, kelly, kelly bushing, master bushing, and rotary table.

rotating head *n:* a sealing device used to close off the annular space around the kelly in drilling with pressure at the surface, usually installed above the main blowout preventers. A rotating head makes it possible to drill ahead even when there is pressure in the annulus that the weight of the drilling fluid is not overcoming; the head prevents the well from blowing out. It is used mainly in the drilling of formations that have low permeability. The rate of penetration through such formations is usually rapid.

rotation gas lift *n:* a gas-lift system in which the gas that is injected and subsequently produced is recompressed and reinjected into the well, effecting a continuous, closed system that does not require the introduction of additional gas from an extraneous source for operation, except that needed to make up losses in the system.

rotor *n:* 1. a device with vanelike blades attached to a shaft; the device turns or rotates when the vanes are struck by a fluid such as drilling mud directed there by a stator. 2. the rotating part of an induction-type alternating current electric motor. Compare *stator.*

roughneck *n:* also called a rotary helper. See *rotary helper.*

round trip *n:* the action of pulling out and subsequently running back into the hole a string of drill pipe or tubing. Making a round trip is also called tripping.

roustabout *n:* 1. a worker on an offshore rig who handles the equipment and supplies that are sent to the rig from the shore base. The head roustabout is very often the crane operator. 2. a worker who assists the foreman in the general work around a producing oilwell, usually on the property of the oil company. 3. a helper on a well servicing unit.

royalty *n:* the part of oil, gas, and minerals or their cash value paid by the lessee to the lessor or to one who has acquired possession of the royalty rights, based on a certain percentage of the gross production from the property.

RP *abbr:* rock pressure; used in drilling reports.

rpm *abbr:* revolutions per minute.

RT *abbr:* rotary table; used in drilling reports.

RUCT *abbr:* rigging up cable tools; used in drilling reports.

run *n:* the amount of crude oil sold and transferred to the pipeline by the producer.

runaround *n:* a platform encircling the top of the derrick.

run a tank *v:* to transfer oil from a stock tank into a pipeline.

run casing *v:* to lower a string of casing into the hole. Also called run pipe.

run in *v:* to go into the hole with tubing, drill pipe, and so forth.

run pipe *v:* to lower a string of casing into the hole.

run ticket *n:* a record of the oil transferred from the producer's storage tank to the pipeline. It is the basic legal instrument by which the lease operator is paid for oil produced and sold.

RUR *abbr:* rigging up rotary rig; used in drilling reports.

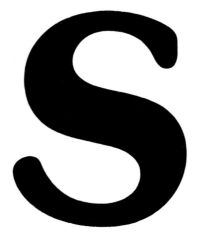

safety clamp *n:* a device used to suspend a rod string after the pump has been spaced or when the weight of the rod string must be taken off the pumping equipment.

safety factor of wire rope *n:* a measurement of load safety for wire rope obtained by using the following formula:

Factor of safety = B/W

where
- B = nominal catalog breaking strength of the wire rope, and
- W = calculated total static load.

Also called design factor.

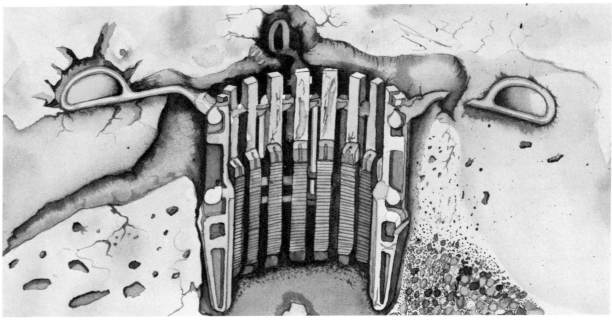
Slips

s *sym:* second.

S *sym:* sulfur.

sack *n:* a container for cement, bentonite, ilmenite, barite, caustic, and so forth. Sacks (bags) contain the following amounts:

Cement	94 lb (1 cu ft)
Bentonite	100 lb
Ilmenite	100 lb
Barite	100 lb

sacrificial anode *n:* in cathodic protection, anodes made from metals whose galvanic potentials render them anodic to steel in an electrolyte. They are used up, or sacrificed.

saddle *n:* See *pipe saddle.*

safety goggles *n:* a protective eye covering worn by oil field workers to minimize the danger to the eyes of being struck by flying objects or harmed by corrosive substances.

safety hat *n:* See *hard hat.*

safety joint *n:* an accessory to the fishing tool, placed above it. If the tool cannot be disengaged from the fish, the safety joint permits easy disengagement of the string of pipe above the safety joint. Thus, part of the safety joint, as well as the tool attached to the fish, remains in the hole and becomes part of the fish.

safety latch *n:* a latch provided on a hook or swivel to prevent it from becoming detached prematurely.

safety platform *n:* the monkeyboard, or platform on a derrick or mast on which the derrickman works. He wears a safety harness (attached to the mast or derrick) to prevent him from falling.

safety shoes *n:* metal-toed shoes or boots with nonskid, corrosion-resistant soles worn by oil field workers to minimize falls and injury to their feet.

safety slide *n:* a device normally mounted near the safety platform to afford the derrickman a means of quick exit to the surface in case an emergency arises. It is usually affixed to a wireline, one end of which is attached to the derrick and the other end to the surface. To exit by the safety slide, the derrickman grasps a handle on it and rides it down to the ground. Also called a geronimo.

safety valve *n:* 1. an automatic valve that opens or closes when an abnormal condition occurs (e.g., a pressure relief valve on a separator that opens if the pressure exceeds the set point, or the shutdown valve at the wellhead that closes if the line pressure becomes too high or too low). 2. a valve installed at the top of the drill stem to prevent flow out of the drill pipe if a kick occurs during tripping operations.

sagging *n:* the distortion of the hull of a vessel when the middle is lower than either end because of excessively heavy or unbalanced loads; the opposite of hogging.

saline drilling fluid *n:* also called salt mud. See *salt mud*.

salinity log *n:* a special radioactivity well log that is electronically adjusted to reflect gamma ray emissions resulting from the collision of neutrons with chlorine atoms in the formations. Salinity, or chlorine, logs provide an estimate of the relative amounts of oil, gas, or salt water in a formation.

salt *n:* a compound that is formed (along with water) by the reaction of an acid with a base. A common salt (table salt) is sodium chloride, NaCl, derived by combining hydrochloric acid, HCl, with sodium hydroxide, NaOH. The result is sodium chloride and water, H_2O. This process is written chemically as

$$HCl + NaOH \rightarrow NaCl + H_2O.$$

Another salt, for example, is calcium sulfate, $CaSO_4$, obtained when sulfuric acid, H_2SO_4, is combined with calcium hydroxide, $Ca(OH)_2$.

salt-brine cement *n:* a cementing slurry whose liquid phase contains sodium chloride.

salt dome *n:* a dome that is caused by an intrusion of rock salt into overlying sediments. A piercement salt dome is one that has been pushed up so that it penetrates the overlying sediments, leaving them truncated. The formations above the salt plug are usually arched so that they dip in all directions away from the center of the dome, thus frequently forming traps for petroleum accumulations.

salt mud *n:* 1. a drilling mud in which the water has an appreciable amount of salt (usually sodium chloride) dissolved in it. Also called saltwater mud or saline drilling fluid. 2. a mud with a resistivity less than or equal to the formation water resistivity.

salt water *n:* a water that contains a large quantity of salt; brine.

saltwater disposal *n:* the method and system for the disposal of salt water produced with crude oil. A typical system is composed of collection centers (in which salt water from several wells is gathered), a central treating plant (in which salt water is conditioned to remove scale- or corrosion-forming substances), and disposal wells (in which treated salt waste is injected into a suitable formation).

saltwater mud *n:* also called salt mud. See *salt mud*.

sample log *n:* a graphic representative model of the rock formations penetrated by drilling, prepared by the geologist from samples and cores.

sampler *n:* a device attached to pipeline to permit continuous sampling of the oil, gas, or product flowing in the line.

samples *n pl:* 1. the well cuttings obtained at designated footage intervals during drilling. From an examination of these cuttings, the geologist determines the type of rock and formations being drilled and estimates oil and gas content. 2. small quantities of well fluids obtained for analysis.

sampling *n:* the taking of a representative sample of fluid from a tank to measure its temperature, specific gravity, and BS&W content.

Samson post *n:* 1. the part of the surface equipment of a standard cable-tool drilling rig that supports the walking beam. 2. the member of a rod pumping unit that supports the walking beam.

sand *n:* 1. an abrasive material composed of small quartz grains formed from the disintegration of preexisting rocks. Sand consists of particles less than 2 mm and greater than $1/16$ mm in diameter. 2. sandstone.

sand consolidation *n:* any one of several methods by which the loose, unconsolidated grains of a producing formation are made to adhere to prevent a well from producing sand but permit it to produce oil and gas.

sand control n: any method by which large amounts of sand in a sandy formation are prevented from entering the wellbore. Sand in the wellbore can cause plugging and premature wear of well equipment. See *gravel pack, screen liner,* and *sand consolidation.*

sanded-up adj: 1. of a well, under restricted production because of sand accumulation in the wellbore. 2. impeded or hindered, especially because of sand accumulation.

sand fill n: a column of sand that has entered and accumulated in the wellbore.

sand lens n: See *lens.*

sand line n: a wireline used on drilling rigs and well-servicing rigs to operate a swab or bailer, to retrieve cores, or to run logging devices. It is usually 9/16 of an inch (15 mm) in diameter and several thousand feet (metres) long.

sand out v: to plug a well inadvertently with proppants during formation fracturing. Sanding out is usually the result of a slowed fracture-fluid velocity, or screening effect, that allows the proppants to become separated from the fluid instead of being carried away from the wellbore. Also called screening out.

sand reel n: a metal drum on a drilling rig or a workover unit around which the sand line is wound. On a drilling rig, it may be attached to the catshaft and may be used for coring or other wireline operations. When used on a drilling rig, it is often called a coring reel.

sandstone n: a detrital sedimentary rock composed of individual grains of sand (commonly quartz) that are cemented together by silica, calcium carbonate, iron oxide, and so forth. Sandstone is a common rock in which petroleum and water accumulate.

sand-thickness map n: a map that shows the thickness of subsurface sands. See *isopachous map.*

S&W abbr: sediment and water.

sat abbr: saturated or saturation; used in drilling reports.

satellite system n: a system that is located some distance from the plant for which it performs a function. Examples are absorbers or compressors.

saturated compounds n pl: hydrocarbon compounds having essentially no unsaturated carbon valence bonds. Natural gas and natural gas liquids are saturated compounds.

saturated liquid n: liquid that is at its boiling point or is in equilibrium with a vapor phase in its containing vessel.

saturated steam n: steam that exists at a temperature corresponding to its absolute pressure. Saturated steam may contain, or be free of, water particles.

saturated vapor n: vapor at its dew point.

saturation n: a state of being filled or permeated to capacity. Sometimes used to mean the degree or percentage of saturation (as, the saturation of the pore space in a formation or the saturation of gas in a liquid, both in reality meaning the extent of saturation).

saturation diving n: diving in which a diver's tissues are saturated with an inert gas to a point where no more of the gas can be absorbed by his body. Consequently, once a diver is saturated, decompression time remains the same whether he stays at the saturated depth for 24 hours or for several days.

saver sub n: a device made up in the drill stem to absorb much of the wear between frequently broken joints (as between the kelly and the drill pipe). See *kelly saver sub.*

Saybolt Seconds Universal n pl: units for measuring the viscosity of lighter petroleum products and lubricating oils. See *Saybolt viscometer.*

Saybolt viscometer n: an instrument used to measure the viscosity of fluids, consisting basically of a container with a hole or jet of a standard size in the bottom. The time required for the flow of a specific volume of fluid is recorded in seconds at three different temperatures (100°F, 130°F, and 210°F). The time measurement unit is referred to as the Saybolt Second Universal (SSU).

scaling tool n: a circular-shaped wire brush that is attached to a pneumatically, hydraulically, or electrically operated tool (such as a grinder) and that is used to remove rust or scale from pipe or other oil field equipment.

scantlings n pl: (nautical) the dimensions of the structural members in the hull.

scavenge v: to remove exhaust gases from a cylinder by means of compressed air. Such removal takes place in all two-cycle diesel engines.

scf abbr: standard cubic feet.

scf/d abbr: standard cubic feet per day.

Schlumberger n: one of the pioneer companies in electric well logging, named for the French scientist who first developed the method; pronounced "slumberjay." Today, many companies provide logging services of all kinds.

scintillation detector *n:* one of four types of detectors used since the inception of radiation logging. Scintillation detection converts tiny flashes of light produced by gamma rays as they expend themselves in certain crystals into electrical pulses. Pulse size depends on the amount of energy absorbed.

scope *n:* the ratio of the total length of a mooring line (as on a mobile offshore drilling rig) to the depth of the water.

SCR *abbr:* silicon controlled rectifier.

scraper *n:* any device that is used to remove deposits (such as scale or paraffin) from tubing, casing, rods, flow lines, or pipelines.

scraper trap *n:* a specially designed piece of equipment that is installed in a pipeline to launch or receive a pipeline scraper.

scratcher *n:* a device that is fastened to the outside of casing and that removes mud cake from the wall of a hole to condition the hole for cementing. By rotating or moving the casing string up and down as it is being run into the hole, the scratcher, formed of stiff wire, removes the cake so that the cement can bond solidly to the formation.

screen liner *n:* a pipe that is perforated and arranged with a wire wrapping to act as a sieve to prevent or minimize the entry of sand particles into the wellbore. Also called a screen pipe.

screen out *v:* See *sand out*.

screen pipe *n:* also called a screen liner. See *screen liner*.

screw packer *n:* a packer in which the packing element is expanded by rotating the pipe; used when it is not desirable to put tubing weight on the packer.

scrubber *n:* 1. a vessel through which fluids are passed to remove dirt and other foreign matter. 2. a vessel with or without internals used to separate entrained liquids or solids from gas. It may be used to protect downstream rotating equipment or to recover valuable liquids from a gas or vapor originating upstream. 3. a unit that removes carbon dioxide from the breathing medium of a diver by chemical absorption.

scrubbing *n:* 1. the purification of gas by treatment in a water or chemical wash. Scrubbing also removes water in the gas. 2. friction wear.

scuba *n:* self-contained underwater breathing apparatus.

sd *abbr:* sand or sandstone; used in drilling reports.

SDO *abbr:* shut down for orders; used in drilling reports.

sdy *abbr:* sandy; used in drilling reports.

seafloor *n:* the bottom of the ocean; the seabed.

seal *v:* to close off or secure against a flow of fluid.

Seale design *n:* a popular design for wire rope for drilling, in which the number of inner wires in each strand is the same as the number of outer wires, but the diameter of the inner wires is smaller than that of the outer.

seal-off *n:* the penetration of a drilling fluid into a potentially productive formation, thus restricting or preventing the formation from producing.

seamless drill pipe *n:* drill pipe that is manufactured in one continuous piece.

sea suctions *n pl:* valve-controlled pipelines, fitted with pumps, which permit a vessel to take on seawater for ballast into any of the vessel's tanks.

seat *n:* the point in the wellbore at which the bottom of the casing is set.

seating nipple *n:* a special tube installed in a string of tubing, having machined contours to fit a matching plug with locking pawls. It is used to hold a regulator, choke, or safety valve; to anchor a pump; or to permit installation of gas-lift valves.

Sec *abbr:* section; used in drilling reports.

second *n:* 1. the fundamental unit of time in the metric system. The symbol for second is s. 2. the crew member who relieves the toolpusher, or rig manager, and is second in command in some countries.

secondary cementing *n:* any cementing operation after the primary cementing operation. Secondary cementing includes a plug-back job, in which a plug of cement is positioned at a specific point in the well and allowed to set. Wells are plugged to shut off bottom water or to reduce the depth of the well for other reasons.

secondary porosity *n:* porosity created in a formation after it has formed, either because of dissolution or stress distortion taking place naturally, or because of treatment by acid or injection of coarse sand.

secondary recovery *n:* 1. pressure maintenance methods used to boost primary production. 2. waterflooding of a depleted reservoir. 3. the first enhanced recovery method of any type applied to a

section milling-set pipe

reservoir to produce oil not recoverable by primary recovery methods. See *pressure maintenance, enhanced oil recovery,* and *primary recovery.*

section milling *n:* the process by which a portion of pipe, usually casing, is actually removed by cutting with a mill.

sed *abbr:* sediment; used in drilling reports.

sediment *n:* 1. the matter that settles to the bottom of a liquid; also called tank bottoms, basic sediment, and so forth. 2. in geology, buried layers of sedimentary rocks.

sediment and water *n:* a material coexisting with, yet foreign to, petroleum liquid and requiring a separate measurement for reasons that include sales accounting. This foreign material includes free water and sediment (dynamic measurement) and/or emulsified or suspended water and sediment (static measurement). The quantity of suspended material present is determined by a centrifuge or laboratory testing of a sample of petroleum liquid.

sedimentary rock *n:* a rock composed of materials that were transported to their present position by wind or water. Sandstone, shale, and limestone are sedimentary rocks.

sedimentation *n:* the process of deposition of layers of rock or sand as material settles out of water, ice, or other material.

seep *n:* the surface appearance of oil or gas that results naturally when a reservoir rock becomes exposed to the surface, thus allowing oil or gas to flow out of fissures in the rock.

segregation drive *n:* See *gravity drainage.*

seis *abbr:* seismograph; used in drilling reports.

seismic data *n:* detailed information obtained from earth vibration produced naturally or artificially (as in geophysical prospecting).

seismograph *n:* a device that detects vibrations in the earth, used in prospecting for probable oil-bearing structures. Vibrations are created by discharging explosives in shallow boreholes or by striking the surface with a heavy blow. The type and velocity of the vibrations as recorded by the seismograph indicate the general characteristics of the section of earth through which the vibrations pass.

seize *v:* to bind the end of a wire rope with fine wire or a metal band to prevent it from unraveling.

self-elevating drilling unit *n:* an offshore drilling rig, usually with a large hull. It has a mat or legs that are lowered to the seafloor and a main deck that is raised above the surface of the water to a distance where it will not be affected by the waves. Also called a jackup drilling rig.

self-potential *n:* also called spontaneous potential. See *spontaneous potential.*

self-potential curve *n:* also called spontaneous potential curve. See *spontaneous potential curve.*

semiclosed circuit *n:* a diving life-support system in which the gas is partially vented and the remainder is recycled, purified, and reoxygenated.

semiexpendable gun *n:* a perforating gun that consists of a metallic strip on which encapsulated shaped charges are mounted. After the gun is fired, the strip is retrieved. See *gun-perforate.*

semisubmersible *n:* See *semisubmersible drilling rig.*

semisubmersible drilling rig *n:* a floating offshore drilling structure that has hulls submerged in the water but not resting on the seafloor. Living quarters, storage space, and so forth are assembled on the deck. Semisubmersible rigs are either self-propelled or towed to a drilling site and either anchored or dynamically positioned over the site or both. Semisubmersibles are more stable than drill ships and are used extensively to drill wildcat wells in rough waters such as the North Sea.

senior orifice fitting *n:* a one-piece orifice fitting that allows the orifice plate in it to be changed without the flow of gas in the line being disturbed.

separator *n:* a cylindrical or spherical vessel used to isolate the components in streams of mixed fluids. See *oil and gas separator.*

sequestering agent *n:* a chemical used with an acid in well treatment to inhibit the precipitation of insoluble iron hydroxides, which form when the acid contacts scales or iron salts and oxides, such as found in corrosion products on casing.

serpentine *n:* an igneous rock composed in part of hydrated magnesium silicate. Hydrocarbons may be associated with serpentine, but not usually.

service company *n:* a company that provides a specialized service, such as a well-logging service or a directional-drilling service.

set back *v:* to place stands of drill pipe and drill collars in a vertical position to one side of the rotary table in the derrick or mast of a drilling or workover rig.

set casing *v:* to run and cement casing at a certain depth in the wellbore. Sometimes, the term *set pipe* is used in reference to setting casing.

set pipe *v:* also called set casing. See *set casing.*

settled production *n:* oil and gas production from long-time fields that produce at approximately the same rate every day.

settling *n:* the separation of substances because of different sizes and specific gravities of components in the substances.

settling pit *n:* the mud pit that is dug in the earth for the purpose of receiving mud returned from the well and allowing the solids in the mud to settle out. Steel mud tanks are more often used today, along with various auxiliary equipment for controlling solids fast and efficiently.

settling tank *n:* the steel mud tank in which solid material in mud is allowed to settle out by gravity; used only in special situations today, for solids control equipment has superseded such a tank in most cases. Sometimes called a settling pit.

set up *v:* to harden (as cement).

sewage treatment plant *n:* a system on offshore locations used to render human and other wastes biologically inert before the wastes are discharged overboard.

SG *abbr:* show of gas; used in drilling reports.

SGA *abbr:* Southern Gas Association.

sh *abbr:* shale; used in drilling reports.

shake out *v:* to spin a sample of oil at high speed, usually in a centrifuge, to determine its BS&W content. Sediments settle out because of the centrifugal force.

shaker *n:* shortened form of shale shaker. See *shale shaker.*

shaker pit *n:* also called shaker tank. See *shaker tank.*

shaker tank *n:* the mud tank adjacent to the shale shaker, usually the first tank into which mud flows after returning from the hole. Also called a shaker pit.

shale *n:* a fine-grained sedimentary rock composed of consolidated silt and clay or mud. Shale is the most frequently occurring sedimentary rock.

shale oil *n:* See *oil shale.*

shale shaker *n:* a vibrating screen used to remove cuttings from the circulating fluid in rotary drilling operations. The size of the openings in the screen should be carefully selected to be the smallest size possible that will allow 100 percent flow of the fluid. Also called a shaker.

shaped charge *n:* a relatively small container of high explosive that is loaded into a perforating gun. Upon detonation, the charge releases a small, high-velocity stream of particles (a jet) that penetrates the casing, cement, and formation. See *gun-perforate.*

shear *n:* action or stress that results from applied forces and that causes or tends to cause two adjoining parts of a body to slide relative to each other in a direction parallel to their plane of contact.

shearometer *n:* an instrument used to measure the shear strength, or gel strength, of a drilling fluid. See *gel strength.*

shear pin *n:* a pin that is inserted at a critical point in a mechanism and that is designed to shear when subjected to excess stress. Usually, the shear pin is easy to replace.

shear ram *n:* the components in a blowout preventer that cut, or shear, through drill pipe and form a seal against well pressure. Shear rams are used in mobile offshore drilling operations to provide a quick method of moving the rig away from the hole when there is no time to trip the drill stem out of the hole.

shear ram preventer *n:* a blowout preventer that uses shear rams as closing elements. See *shear ram.*

shear relief valve *n:* a type of pressure relief valve in which excess pressure causes a shearing action on a pin to relieve pressure. See *pressure relief valve.*

shear strength *n:* See *gel strength.*

shear thinning *n:* viscosity reduction of non-Newtonian fluids (e.g., polymers, most slurries and suspensions, and lube oils with viscosity-index improvers) under conditions of shear stress.

sheave *n:* (pronounced "shiv") a grooved pulley.

shell *n:* 1. the body of a tank. 2. the horizontal tank on a tank car, containing the liquid being transported. 3. the steel backing of a precision insert bearing on a bit.

Shepard's cane *n:* an earth-resistivity meter used to measure the resistance of soil to the passage of electrical current.

shim *n:* a thin, often tapered piece of material used to fill in space between things (as for support, leveling, or adjustment of fit).

shipshape drilling rig *n:* drill ship.

shirttail *n:* the part of a drilling bit on which the cone is anchored. Shirttails extend below the threaded pin of the bit and are usually rounded on bottom, thus acquiring the name.

shoe *n:* also called guide shoe or casing hose. See *guide shoe*.

shoestring sand *n:* a long, narrow sand deposit, usually a buried sandbar or filled channel.

shoot *v:* 1. to explode nitroglycerine or other high explosives in a hole to shatter the rock and increase the flow of oil; now largely replaced by formation fracturing. 2. in seismographic work, to discharge explosives to create vibrations in the earth's crust. See *seismograph*.

shot *n:* 1. a charge of high explosive, usually nitroglycerine, detonated in a well to shatter the formation and expedite the recovery of oil. Shooting has been almost completely replaced by formation fracturing and acid treatments. 2. a point at which a photograph is made in a single-shot survey. See *directional survey*.

shot-hole drilling *n:* the drilling of relatively small holes into the earth, the purpose of which is to provide a means for lowering an explosive shot in order to create shock waves for seismic analysis.

shoulder *n:* 1. the flat portion, machined on the base of the bit shank, that meets the shoulder of the drill collar and serves to form a pressure-tight seal between the bit and the drill collar. 2. the flat portion of the box end and of the pin end of a tool joint; the two shoulders meet when the tool joint is connected and form a pressure-tight seal.

show *n:* the appearance of oil or gas in cuttings, samples, or cores from a drilling well.

shrinkage *n:* 1. a decrease in oil volume caused by evaporation of solution gas or by lowered temperature. 2. the reduction in volume or heating value of a gas stream due to removal of some of its constituents. 3. the unaccounted loss of products from storage tanks.

shrink-on tool joint *n:* a tool joint made to fit the pipe by the process of shrinking on, that is, by heating the outer member to expand the bore for easy assembly and then cooling it so that it contracts around the inner member.

shrouded jet nozzle *n:* a special type of jet nozzle that is manufactured with a projection (the shroud), which serves to minimize the erosion of the nozzle by the high-velocity jet of drilling fluid being forced through it.

shunt *n:* a conductor joining two points in an electrical circuit to form a parallel or alternative path through which a portion of the current may pass.

shut down *v:* to stop work temporarily or to stop a machine or operation.

shutdown rate *n:* a rate provision that is usually contained in a drilling contract and that specifies the compensation to the independent drilling contractor when drilling is suspended at the request of the operator.

shut in *v:* 1. to close the valves on a well so that it stops producing. 2. to close in a well in which a kick has occurred.

shut-in *adj:* shut off to prevent flow. Said of a well, plant, pump, and so forth, when valves are closed at both inlet and outlet.

shut-in bottomhole pressure *n:* the pressure at the bottom of a well when the surface valves on the well are completely closed. The pressure is caused by fluids that exist in the formation at the bottom of the well.

shut-in casing pressure *n:* pressure of the annular fluid on the casing when a well is shut in.

shut-in drill pipe pressure *n:* pressure of the drilling fluid on the inside of the drill stem; used to measure the difference between hydrostatic pressure and formation pressure when a well is shut in after a kick and the mud pump is off.

shut-in pressure *n:* the pressure when the well is completely shut in, as noted on a gauge installed on the surface control valves. When drilling is in progress, shut-in pressure should be zero, because the pressure exerted by the drilling fluid should be equal to or greater than the pressure exerted by the formations through which the wellbore passes. On a flowing, producing well, however, shut-in pressure should be above zero.

shut off *v:* to stop or decrease the production of water in an oilwell by cementing or mudding off the water-producing interval.

shuttle vessel *n:* a tank ship capable of navigating shallow ports, usually 30,000–70,000 dwt. It shuttles between anchored large vessels and port.

SI *abbr:* 1. shut in; used in drilling reports. 2. Système International. See *international system of units*.

SIBHP *abbr:* shut-in bottomhole pressure; used in drilling reports.

SICP *abbr:* shut-in casing pressure.

side-door mandrel *n:* See *gas-lift mandrel*.

side-pocket mandrel *n:* See *gas-lift mandrel*.

sidetrack *v:* to drill around broken drill pipe or casing that has become lodged permanently in the hole, using a whipstock, turbodrill, or other mud motor.

sidewall coring *n:* a coring technique in which core samples are obtained from a zone that has already been drilled. A hollow bullet is fired into the formation wall to capture the core and then retrieved on a flexible steel cable. Core samples of this type usually range from ¾ to 1³/₁₆ inches (20 to 30 mm) in diameter and from ¾ to 4 inches (20 to 100 mm) in length. This method is especially useful in soft-rock areas.

sidewall epithermal neutron log *n:* a device in which a neutron source and a detector are in a pad using the same hardware as the density log. The pad is pushed against the side of the borehole wall.

SIDPP *abbr:* shut-in drill pipe pressure; used in drilling reports.

sieve analysis *n:* the determination of the percentage of particles that pass through several screens of graduated fineness.

sieve tray *n:* a tray installed in an absorber tower or fractionating column similar to a bubble cap tray except that the tray has only holes but no bubble caps through it. This type of tray is more efficient than the bubble cap tray or the valve tray and less expensive than either. However, it does not operate properly over a wide range of flow rates.

silica flour *n:* a silica (SiO_2) ground to a fineness equal to that of portland cement.

silicon controlled rectifier *n:* a device that changes alternating current to direct current by means of a silicon control gate. Commonly called SCR or Thyristor.

single *n:* a joint of drill pipe. Compare *double*, *thribble*, and *fourble*.

single-pole rig *n:* a well-servicing unit whose mast consists of but one steel tube, usually about 65 feet long.

single-shot survey *n:* a directional survey that provides a single record of the drift direction and off-vertical orientation of the hole. See *directional survey*.

sinker bar *n:* a heavy weight or bar placed on or near a lightweight wireline tool. The bar provides weight so that the tool will lower properly into the well.

sinter *v:* to bond metallic powder into a mass by heating it. Tungsten carbide inserts are often bonded to the cones of button bits by sintering when the bits are being manufactured.

SIP *abbr:* shut-in pressure; used in drilling reports.

siphon *n:* a bent tube or pipe through which a liquid flows, first rising to a higher level than that of the pipe inlet and then flowing downward to a lower level than the inlet.

skid the rig *v:* to move a rig with a standard derrick from the location of a lost or completed hole preparatory to starting a new hole. Skidding the rig allows the move to be accomplished with little or no dismantling of equipment.

skim pit *n:* an earthen pit, often lined with concrete, into which water with small amounts of oil is pumped. The minute quantities of oil are skimmed off the top of the water in the pit, and the water is disposed of.

skin *n:* 1. the area of the formation that is damaged because of the invasion of foreign substances into the exposed section of the formation adjacent to the wellbore during drilling and completion. 2. the pressure drop from the outer limits of drainage to the wellbore caused by the relatively thin veneer (or skin) of the affected formation. Skin is expressed in dimensionless units; a positive value denotes formation damage, and a negative value indicates improvement.

skin effect *n:* See *skin*.

sky-top mast *n:* a mast on a well servicing unit that utilizes a split traveling block and crown block, which makes it possible to pull 60-foot stands with a 50-foot mast.

slack off *v:* to lower a load or ease up on a line. A driller will slack off to put additional weight on the bit.

slaked lime *n:* See *hydrated lime*.

sleeve *n:* a tubular part designed to fit over another part.

slick line *n:* also called wireline. See *wireline*.

sliding-sleeve nipple *n:* a special device placed in a string of tubing and operated by a wireline tool (1) to open or close orifices, thus permitting circulation between the tubing and annulus, or (2) to open or shut off production from alternate intervals in a well.

slim-hole drilling *n:* drilling in which the size of the hole is smaller than the conventional hole diameter for a given depth. This decrease in hole size enables the operator to run smaller casing, thereby lessening the cost of completion. See *miniaturized completion*.

slip *v:* to move drilling line periodically so that it wears evenly as it is used.

slip elevator *n:* a casing elevator containing segmented slips with gripping teeth inside. Slip

slip joint–snub

elevators are recommended for long strings of casing because the teeth grip the casing and help prevent casing damage from the weight of long, heavy strings hanging from elevators. Slip elevators may also be used as slips.

slip joint n: See *telescoping joint*.

slipping and cutoff program n: procedure for a given rig in which drilling line is slipped through the system at such a rate that it is evenly worn, then cut off at the drum end just as it reaches the end of its useful life.

slip ring n: a conducting ring that gives current to or receives current from the brushes in a generator or motor.

slips n pl: wedge-shaped pieces of metal with teeth or other gripping elements that are used to prevent pipe from slipping down into the hole or to hold pipe in place. Rotary slips fit around the drill pipe and wedge against the master bushing to support the pipe. Power slips are pneumatically or hydraulically actuated devices that allow the crew to dispense with the manual handling of slips when making a connection. Packers and other downhole equipment are secured in position by slips that engage the pipe by action directed at the surface.

slop n: a term rather loosely used to denote mixtures of oil produced at various places in a plant and requiring rerun or other processing to be suitable for use. Also called slop oil.

slops n: oil that has been washed from the tanks of a vessel and is pumped to a special tank where most of the water will be permitted to separate for decanting.

sloughing n: (pronounced "sluffing"). Also called caving. See *caving*.

slow-release inhibitor n: corrosion-preventive substance that is released into production fluids at a slow rate.

slow-set cement n: a manufactured cement in which the thickening time is extended by the use of a coarser grind, the elimination of the rapid hydrating components in its composition, and the addition of a chemical retarder. API classes N, D, E, and F are slow-set cements.

sludge n: a tarlike substance that is formed when oil oxidizes.

slugging compound n: a special chemical demulsifier that is often added to the emulsion samples to determine the total amount of sediment and water in the samples; also called knockout drops.

slug the pipe v: to pump a quantity of heavy mud into the drill pipe. Before hoisting drill pipe, it is desirable (if possible) to pump into its top section a quantity of heavy mud, or a slug, that causes the level of the fluid to remain below the rig floor so that the crew members and the rig floor are not contaminated with the fluid when stands are broken out.

slurry n: a plastic mixture of cement and water that is pumped into a well to harden; there it supports the casing and provides a seal in the wellbore to prevent migration of underground fluids.

slurry viscosity n: the consistency of a slurry, measured in poise.

slurry volume n: the sum of the absolute volumes of solids and liquids that constitute a slurry.

slurry weight n: the density of a cement slurry, expressed in pounds per gallon (ppg), pounds per cubic feet (lb/ft^3), kilograms per litre (kg/L), etc.

slurry yield n: the volume of slurry obtained when one sack of cement is mixed with the desired amount of water and additives (as with accelerators, fluid-loss control agents, etc.)

slush pit n: the old term for a mud pit. See *mud pit*.

slush pump n: also called mud pump. See *mud pump*.

snake n: also called a swivel-connector grip. See *swivel-connector grip*.

snatch block n: 1. a block that can be opened to receive wire rope or wireline. 2. a block that is suited for a single sheave and is used for pulling horizontally on an A-frame mast.

SNG abbr: synthetic or substitute natural gas.

sniffer n: See *explosimeter*.

snipe n: also called a cheater. See *cheater*.

snub v: 1. to force pipe or tools into a high-pressure well that has not been killed (i.e., to run pipe or tools into the well against pressure when the weights of pipe are not great enough to force the pipe through the BOPs). Snubbing usually requires an array of wireline blocks and wire rope that forces the pipe or tools into the well through a stripper head or blowout preventer until the weight of the string is sufficient to overcome the lifting effect of the well pressure on the pipe in the stripper. In workover operations, snubbing is usually accomplished by using hydraulic power to force the pipe through the stripping head or blowout preventer. 2. to tie up short with a line.

snubber *n:* 1. a device that hydraulically forces pipe or tools into the well against pressure. 2. a device within some hooks that acts as a shock absorber in eliminating the bouncing action of pipe as it is picked up.

snubbing line *n:* 1. a line used to check or restrain an object. 2. a wire rope used to put pipe or tools into a well while the well is closed in. See *snub*.

snuffer *n:* a tank safety device that seals the vapor vent manually and prevents vapors from escaping into a fire, thus snuffing out the flame.

SO *abbr:* show of oil; used in drilling reports.

SO$_2$ *form:* sulfur dioxide.

SO&G *abbr:* show of oil and gas; used in drilling reports.

socket *n:* 1. a hollow object or open device that fits or holds an object. 2. any of several fishing tools used to grip the outside of a lost tool or a joint of pipe.

sodium carboxymethyl cellulose *n:* See *carboxymethyl cellulose*.

sodium chloride *n:* common table salt; sometimes used in cement slurries as an accelerator or a retarder, depending on the concentration. Chemical formula is NaCl.

soft crossover system *n:* a pattern of drum spooling in which the wire rope travels in a two-step grooving pattern but has flat or level areas for crossing over to act as shock absorbers for the rope and to reduce the rise or hump produced in all multiwrapping of wire rope.

soft water *n:* See *hard water*.

soil stress *n:* the uneven penetration of pipeline coatings due to changes in soil volume and moisture along the pipeline bed.

solenoid *n:* a cylindrical coil of wire that resembles a bar magnet when it carries a current so that it draws a movable core into the coil when the current flows.

solid wireline *n:* a special wireline made of brittle but very strong steel, usually 0.066 to 0.092 inches in diameter (as opposed to stranded wirelines, which may be $3/16$ inch or larger). Solid, or slick, wirelines are used in depth measurements and in running special devices into a well under pressure.

solution *n:* a single, homogeneous liquid, solid, or gas phase that is a mixture in which the components (liquid, gas, solid, or combinations thereof) are uniformly distributed throughout the mixture. In a solution, the dissolved substance is called the solute; the substance in which the solute is dissolved is called the solvent.

solution gas *n:* lighter hydrocarbons that exist as a liquid under reservoir conditions but that become a gas when the reservoir is produced. See *solution-gas drive*.

solution-gas drive *n:* a source of natural reservoir energy, in which the solution gas coming out of the oil expands to force the oil into the wellbore.

solution gas-oil ratio *n:* See *gas-oil ratio*.

sonde *n:* a logging tool assembly, especially the device in the logging assembly that senses and transmits formation data.

sonic log *n:* a type of acoustic log that records the noise of fluid movement in channels in cement behind casing.

sonic logging *n:* the recording of the time required for a sound wave to travel a specific distance through a formation. Difference in observed travel times is largely caused by variations in porosities of the medium, an important determination. The sonic log, which may be run simultaneously with a spontaneous-potential log or a gamma-ray log, is useful for correlation and often is used in conjunction with other logging services for substantiation of porosities. It is run in an uncased hole.

sour *adj:* containing or caused by hydrogen sulfide or another acid gas (e.g., sour crude, sour gas, sour corrosion).

source-detector spacing *n:* the spacing on a neutron logging device between the neutron source and the detector. Total count rate, porosity resolution, and borehole effects influence source-detector spacing. The spacing is selected specifically for the source-detector characteristics exhibited by each device; optimum spacings must be determined for each tool.

source station *n:* a pump station at a pipeline junction, where oil is pumped from a main line into a branch or lateral line.

sour corrosion *n:* embrittlement and subsequent wearing away of metal, caused by contact of the metal with hydrogen sulfide.

sour crude *n:* also called sour crude oil. See *sour crude oil*.

sour crude oil *n:* oil containing hydrogen sulfide or another acid gas.

sour gas *n:* gas containing an appreciable quantity of hydrogen sulfide and/or mercaptans.

Southern Gas Association n: a Dallas-based organization founded to promote the development of the gas industry, to encourage scientific research affecting the industry, to exchange ideas and information among member companies, and to cooperate with other organizations having mutual objectives. SGA is the largest of four regional organizations started in 1908.

SP abbr: spontaneous potential or self-potential.

spacing n: See *well spacing*.

spacing clamp n: a clamp used to hold the rod string in pumping position when the well is in the final stages of being put back on the pump.

spaghetti n: tubing or pipe with a very small diameter.

spall v: to break off in chips or scales.

spd abbr: spudded; used in drilling reports.

SPE abbr: Society of Petroleum Engineers.

spear n: a fishing tool used to retrieve pipe lost in a well. The spear is lowered down the hole and into the lost pipe, and, when weight, torque, or both are applied to the string to which the spear is attached, the slips in the spear expand and tightly grip the inside of the wall of the lost pipe. Then the string, spear, and lost pipe are pulled to the surface.

spearhead n: See *preflush*.

specific gravity n: the ratio of the weight of a given volume of a substance at a given temperature to the weight of an equal volume of a standard substance at the same temperature. For example, if 1 cubic inch of water at 39°F weighs 1 unit and 1 cubic inch of another solid or liquid at 39° weighs 0.95 unit, then the specific gravity of the substance is 0.95. In determining the specific gravity of gases, the comparison is made with the standard of air or hydrogen. See *gravity*.

specific heat n: the amount of heat required to cause a unit increase in temperature in a unit mass of a substance, expressed as numerically equal to the number of calories needed to raise the temperature of 1 g of a substance by 1°C.

speed droop n: the number of revolutions per minute that an engine slows down from running at maximum no-load speed to running at maximum full-load speed. Usually expressed as a percentage, speed droop should not exceed 7 percent.

speed kit n: a dual-speed traveling block, which permits one elevator to pick up stands as they are broken out while the traveling block continues to move.

spent adj: descriptive of a substance whose strength of merit has been exhausted in a process. For example, after a well has been acidized, any acid that remains in the well is said to be a spent acid because its strength has been used up in the acidizing process.

spider n: a circular steel device that holds slips supporting a suspended string of drill pipe, casing, or tubing. A spider may be split or solid.

spinner survey n: a production-logging method that uses a small propeller turned by fluid movement. By use of a recording arrangement, the number of turns of the propeller can be related to the fluid quantity flowing past the instrument to obtain a production log.

spinning cathead n: also called makeup cathead. See *makeup cathead* and *spinning chain*.

spinning chain n: a chain used to spin up (tighten) one joint of drill pipe into another. In use, one end of the chain is attached to the tongs, another end to the spinning cathead, and the third end left free. The free end is wrapped around the tool joint, and the cathead pulls the chain off the joint, causing the joint to spin (turn) rapidly and tighten up. After the chain is pulled off the joint, the tongs are secured in the same spot, and continued pull on the chain (and thus on the tongs) by the cathead makes up the joint to final tightness.

spinning wrench n: air-powered or hydraulically powered wrench used to spin drill pipe in making or breaking connections.

spin-up n: the rapid turning of the drill stem when one length of pipe is being joined to another.

spirally grooved drill collar n: a drill collar with a round cross section that has a long continuous groove or flute machined helically into its outer surface. The spiraled groove provides space between the wall of the hole and the body of the collar, minimizing the area of contact between the hole wall and the collar; thus the possibility of differential pressure sticking is reduced.

splash box n: See *mother hubbard*.

splash zone n: the area on an offshore structure that is regularly wetted by seawater but is not continuously submerged. Metal in the splash zone must be well protected from the corrosive action of seawater and air.

splice v: to join two parts of a rope or wireline by interweaving individual strands of the line together. Unlike a knot, a splice does not significantly increase the diameter of the line at the point where the parts are joined.

splitter *n:* another term for a fractionator, particularly one that separates isomers. For example, a deisobutanizer may be called a butane splitter.

sponge absorbent *n:* an absorbent for recovering vapors of a lighter absorbent that is used in the main absorption process of a gas processing plant.

sponge absorption unit *n:* a unit wherein the vapors of lighter absorption oils are recovered.

spontaneous combustion *n:* the ignition of combustible materials without open flame.

spontaneous potential *n:* one of the natural electrical characteristics exhibited by a formation as measured by a logging tool lowered into the wellbore. Also referred to as self-potential, it is one of the basic curves obtained by an electrical well log; usually referred to by the initials *SP*.

spontaneous potential curve *n:* a measurement of the electrical currents that occur in the wellbore when fluids of different salinities are in contact. The SP curve is usually recorded in holes drilled with freshwater-base drilling fluids. Also called self-potential curve.

spool *n:* 1. the drawworks drum. 2. also called a drilling spool. See *drilling spool*. 3. also called a casinghead. See *casinghead*. *v:* to wind around a drum.

spot *v:* to pump a designated quantity of a substance (such as acid or cement) into a specific interval in the well. For example, 10 barrels of diesel oil may be spotted around an area in the hole in which drill collars are stuck against the wall of the hole in an effort to free the collars.

spray valve *n:* See *nozzle*.

spread mooring system *n:* a system of rope, chain, or combination of the two attached to anchors on the ocean floor and winches on the structure to keep a floating vessel near a fixed location on the sea surface.

spring collet *n:* a spring-actuated metal band or ring (ferrule) used to expand a liner patch when making casing repairs. See *liner patch*.

sprocket *n:* 1. a wheel with projections on the periphery to engage with the links of a chain. 2. the projection itself that fits into an opening of a chain.

spud *v:* to move the drill stem up and down in the hole over a short distance without rotation. Careless execution of this operation creates pressure surges that can cause a formation to break down and results in lost circulation. See *spud in*.

spud bit *n:* a special kind of drilling bit with sharp blades rather than teeth. It is sometimes used for drilling soft, sticky formations.

spudder *n:* a portable cable-tool drilling rig, sometimes mounted on a truck or trailer.

spud in *v:* to begin drilling; to start the hole.

spur line *n:* an oil pipeline that picks up oil from the gathering lines of several oil fields and delivers it to a main line or trunk line.

sq *abbr:* square.

squ *abbr:* squeeze; used in drilling reports.

square drill collar *n:* a special drill collar, square but with rounded edges, used to control the straightness or direction of the hole; often part of a packed-hole assembly.

square-drive master bushing *n:* a master bushing that has a square opening or recess to accept and drive the square that is on the bottom of the square-drive kelly bushing.

square metre *n:* a unit of metric measure of an area equal to a square that measures 1 m on each side.

squeeze *n:* 1. a cementing operation in which cement is pumped behind the casing under high pressure to recement channeled areas or to block off an uncemented zone. 2. the increasing of external pressure upon a diver's body by improper diving technique.

squeeze cementing *n:* the forcing of cement slurry by pressure to specified points in a well to cause seals at the points of squeeze. It is a secondary cementing method that is used to isolate a producing formation, seal off water, repair casing leaks, and so forth.

squnch joint *n:* a special threadless tool joint for large-diameter pipe, especially conductor pipe, sometimes used on offshore drilling rigs. When the box is brought down over the pin and weight is applied, a locking device is actuated to seat the joints. Because no rotation is required to make up these joints, their use can save time when the conductor pipe is being run.

ss *abbr:* sand or sandstone; used in drilling reports.

SSO *abbr:* slight show of oil; used in drilling reports.

SSTT *abbr:* subsea test tree.

SSU *abbr:* Saybolt seconds universal.

S/T *abbr:* sample tops; used in drilling reports.

stab *v*: to guide the end of a pipe into a coupling or tool joint when making up a connection.

stabbing board *n*: a temporary platform erected in the derrick or mast some 20 to 40 feet (6-12 m) above the derrick floor. The derrickman or another crew member works on the board while casing is being run in a well. The board may be wooden or fabricated of steel girders floored with antiskid material and powered electrically to be raised or lowered to the desired level. A stabbing board serves the same purpose as a monkeyboard but is temporary instead of permanent.

stabbing jack *n*: See *jack board*.

stabbing protector *n*: a protective device, usually made of rubber, that fits on the outside diameter of the box of the pipe that is in the hole; it has a funnel-shaped top and serves as a cushion and guide for stabbing pipe.

stability *n*: the ability of a ship or mobile offshore drilling rig to return to an upright position when it has rolled to either side because of an external force (such as waves).

stabilized condensate *n*: condensate that has been stabilized to a definite vapor pressure in a fractionation system.

stabilizer *n*: 1. a tool placed near the bit, and often just above it, in the drilling assembly and used to change the deviation angle in a well by controlling the location of the contact point between the hole and the drill collars. Conversely, stabilizers are used to maintain correct hole angle. See *packed-hole assembly*. 2. a vessel in which hydrocarbon vapors are separated from liquids. 3. a fractionation system that reduces the vapor pressure so that the resulting liquid is less volatile.

stable emulsion *n*: See *emulsion*.

stack *n*: 1. a vertical pile of blowout prevention equipment. Also called preventer stack. See *blowout preventer*. 2. the vertical chimneylike installation that is the waste disposal system for unwanted vapor such as flue gases or tail-gas streams.

stack a rig *v*: to store a drilling rig upon completion of a job when the rig is to be withdrawn from operation for a time.

stage separation *n*: an operation in which well fluids under pressure are separated into liquid and gaseous components by being passed consecutively through two or more separators. The operating pressure of each succeeding separator is slower than the one preceding it. Stage separation is an efficient process in that a high percentage of the light ends of the fluid is conserved.

staging *n*: the placement of compressors, pumps, cooling systems, treating systems, and so forth, in a series with another unit or units of like design to improve operating efficiency and results.

stake a well *v*: to locate precisely on the surface of the ground the point at which a well is to be drilled. After exploration techniques have revealed the possibility of the existence of a subsurface hydrocarbon-bearing formation, a certified and registered land surveyor drives a stake into the ground to mark the spot where the well is to be drilled.

stand *n*: the connected joints of pipe racked in the derrick or mast during a trip. The usual stand is 90 feet long (about 27 m), which is three lengths of drill pipe screwed together (a thribble).

standard cubic foot *n*: a gas volume unit of measurement at a specified temperature and pressure. The temperature and pressure may be defined in the gas sales contract or by reference to other standards. Its abbreviation is scf.

standard derrick *n*: a derrick that is built piece by piece at the drilling location, as opposed to a jackknife mast, which is preassembled. Standard derricks have been replaced almost totally by jackknife masts.

standard dress *n*: diving equipment consisting of brass diving helmet, breastplate, heavy dry suit, weighted boots, weighted belt, hose, compressor, and communications.

standard gas measurement law *n*: a law, specific for each of several states, that defines the pressure and temperature bases under which a standard cubic foot of gas should be measured in the particular state. The standard applies only to the specific state and may not agree with that in another state.

standard pressure *n*: the pressure exerted by a column of mercury 760 mm high; equivalent to 14.7 psia. Compare *base pressure*.

standard temperature *n*: a predetermined temperature used as a basic measurement. The petroleum industry uses 60°F (15.5°C) as its standard temperature during measurement of oil. The volume of a quantity of oil at its actual temperature (assuming it is not 60°F) is converted to the volume the oil would occupy at 60°F. Conversion is aided by the use of API conversion tables.

standing valve *n*: a fixed ball-and-seat valve at the lower end of the working barrel of a sucker rod pump. The standing valve and its cage do not move as does the traveling value.

standoff *n:* in perforating, the distance a jet or bullet must travel in the wellbore before encountering the wall of the hole.

standpipe *n:* a vertical pipe rising along the side of the derrick or mast, which joins the discharge line leading from the mud pump to the rotary hose and through which mud is pumped going into the hole.

stand tubing *v:* to support tubing in the derrick or mast when it is out of the well rather than to lay it on a rack. Portable workover rigs are usually fitted with a mast that holds stands about 60 feet long (about 18 m), also known as doubles.

starboard *n:* (nautical) the right side of a vessel (determined by looking toward the bow).

starch *n:* a complex carbohydrate sometimes added to drilling fluids to reduce filtration loss.

static fluid level *n:* the level to which fluid rises in a well when the well is shut in.

static pressure *n:* the pressure exerted by a fluid upon a surface that is at rest in relation to the fluid.

stator *n:* 1. a device with vanelike blades that serves to direct a flow of fluid (such as drilling mud) onto another set of blades (called the rotor). The stator does not move; rather, it serves merely to guide the flow of fluid at a suitable angle to the rotor blades. 2. the stationary part of an induction-type alternating-current electric motor. Compare *rotor*.

STB *abbr:* stock tank barrel.

STB/d *abbr:* stock tank barrels per day.

std *abbr:* standard.

stds *abbr:* stands; used in drilling reports.

steam *n:* water in its gaseous state.

steam boiler *n:* a closed steel vessel or container in which water is heated to produce steam.

steam coil *n:* a pipe, or set of pipes, within an emulsion settling tank through which steam is passed to warm the emulsion and make the oil less viscous. See *fire tube*.

steam drive *n:* a method of enhanced oil recovery in which steam is injected into a reservoir through injection wells and driven toward producing wells. The steam reduces the viscosity of crude oil, causing it to flow more freely. The heat vaporizes lighter hydrocarbons; as they move ahead of the steam, they cool and condense into liquids that dissolve and displace crude oil. The steam provides additional gas drive. This method is used to recover viscous oils. Also called continuous steam injection or steam flooding.

steam rig *n:* a rotary drilling rig on which steam engines operate as prime movers. High-pressure steam is furnished by a boiler plant located near the rig. Steam rigs have been replaced almost totally by mechanical or electric rigs.

steel *n:* a malleable alloy of iron and carbon that also contains appreciable amounts of manganese and other elements.

steel-tooth bit *n:* a roller cone bit in which the surface of each cone is made up of rows of steel teeth. Also called a milled-tooth bit or milled bit.

stemming *n:* material used to hold back the force of an explosion, such as sand, gravel, or cement plug placed in a well above a nitroglycerin charge.

step-out well *n:* a well drilled adjacent to or near a proven well to ascertain the limits of the reservoir; an outpost well.

stiff drilling assembly *n:* also called packed-hole assembly. See *packed-hole assembly*.

still *n:* any vessel in which hydrocarbon distillation is effected.

stimulation *n:* any process undertaken to enlarge old channels or create new ones in the producing formation of a well (e.g., acidizing or formation fracturing).

stinger *n:* 1. a cylindrical or tubular projection, relatively small in diameter, that extends below a downhole tool and helps to guide the tool to a designated spot (as into the center of a portion of stuck pipe). 2. a device for guiding pipe and lowering it to the water bottom as it is being laid down by a lay barge.

stn *abbr:* stain; used in drilling reports.

stock tank *n:* a storage tank for treated crude oil. Compare *production tank*.

stock tank oil *n:* oil as it exists at atmospheric conditions in a stock tank. Stocktank oil lacks much of the dissolved gas present at reservoir pressure and temperatures.

stopcock *n:* a valve that shuts off or regulates fluid flow. *v:* to shut in intermittently a producing oilwell to allow a buildup of gas pressure, which effects a more efficient recovery.

storage gas *n:* gas that is stored in an underground reservoir.

storage tank *n:* a tank in which oil is stored pending transfer by pipeline, truck, or other vehicle for selling.

Storm Choke *n:* a proprietary name for a tubing safety valve.

straddle packer *n:* two packers separated by a spacer of variable length. A straddle packer may be used to isolate sections of open hole to be treated or tested or to isolate certain areas of perforated casing from the rest of the perforated section.

straddle test *n:* selective testing of an interval or formation by the use of two packers, one above and one below the zone being tested.

straight hole *n:* a hole that is drilled vertically. The total hole angle is restricted, and the hole does not change direction rapidly – no more than 3° per 100 feet (30.48 m) of hole.

strain *v:* to effect a change of form or size as a result of the application of a stress.

strain gauge *n:* an instrument used to measure minute distortions caused by stress forces in mechanical components.

stranding machine *n:* also called a closing machine. See *closing machine*.

strap *v:* to measure and record the dimensions of oil tanks to prepare tank tables (gauging tables) for determining accurately the volume of oil in a tank at any measured depth.

strap in *v:* to measure a length of pipe as it is run into the hole.

strata *n pl:* distinct, usually parallel beds of rock. An individual bed is a stratum.

stratification *n:* the natural layering or lamination characteristic of sediments and sedimentary rocks.

stratigraphic test *n:* See *strat test*.

stratigraphic trap *n:* a petroleum trap that occurs when the top of the reservoir bed is terminated by other beds or by a change of porosity or permeability within the reservoir itself. Compare *structural trap*.

stratigraphy *n:* a branch of geology concerned with the study of the origin, composition, distribution, and succession of rock strata.

strat test *n:* the drilling of a well primarily to obtain geological information, usually not completed even if commercial quantities of petroleum are found.

stratum *n:* See *strata*.

stray current *n:* a portion of an electric current that flows over a path other than the intended path, causing corrosion of structures immersed in the same electrolyte.

stream *n:* the liquid or gas contained in any pipeline or flowing line.

stream day *n:* a day of full operation by a plant; a basis for calculating plant production. It is different from a calendar day, which would be used to give average production for a full year.

stress *n:* a force that, when applied to an object, distorts or deforms it.

stress concentrator *n:* a notch or pit on a pipe or joint that raises the stress level and concentrates the breakdown of the metal structure. Also called a stress riser.

stress riser *n:* also called a stress concentrator. See *stress concentrator*.

strike *n:* See *formation strike*.

strike plate *n:* an extra piece of metal placed on the bottom of an oil storage tank to protect it from the repeated striking of the plumb bob at the end of the gauger's tape.

string *n:* the entire length of casing, tubing, sucker rods, or drill pipe run into a hole.

stringer *n:* an extra support placed under the middle of racked pipe to keep the pipe from sagging.

string shot *n:* an explosive device that uses primacord, a textile-covered fuse with a core of very high explosive, to create an explosive jar inside stuck pipe or tubing to back off the pipe at the joint immediately above the stuck pipe. See *shot*.

string-shot back-off *n:* See *string shot*.

string up *v:* to thread the drilling line through the sheaves of the crown block and traveling block. One end of the line is secured to the hoisting drum and the other to the derrick substructure.

strip a well *v:* to pull rods and tubing from a well at the same time – for example, when the pump is stuck. Tubing must be stripped over the rods a joint at a time, and the exposed sucker rod is then backed off and removed.

stripped gas *n:* a processed gas from which liquefied hydrocarbons have been removed.

stripper *n:* 1. a well nearing depletion that produces a very small amount of oil or gas. 2. a stripper head. See *stripper head*. 3. a column wherein absorbed constituents are stripped from absorption oil. The term is applicable to columns using stripping medium, such as steam or gas.

stripper head *n:* a blowout prevention device consisting of a gland and packing arrangement bolted to the wellhead. It is often used to seal the annular space between tubing and casing.

stripper rubber *n:* 1. a rubber disk surrounding drill pipe or tubing that removes mud as the pipe is brought out of the hole. 2. the pressure-sealing element of a stripper blowout preventer. See *stripper head.*

stripping in *n:* 1. the process of lowering the drill stem into the wellbore when the well is shut in on a kick. 2. the process of putting tubing into a well under pressure.

stripping job *n:* the simultaneous pulling of rods and tubing when the sucker rod pump or rods are frozen in the tubing string.

stripping out *n:* 1. the process of raising the drill stem out of the wellbore when the well is shut in on a kick. 2. the process of removing tubing from the well under pressure.

strip pipe *v:* 1. to remove the drill stem from the hole while the blowout preventers are closed. 2. to pull the drill stem and the washover pipe out of the hole at the same time.

strks *abbr:* streaks; used in drilling reports.

structural mast *n:* a portable mast constructed of angular as opposed to tubular steel members. See *jackknife mast.*

structural trap *n:* a petroleum trap that is formed because of deformation (as folding or faulting) of the rock layer that contains petroleum. Compare *stratigraphic trap.*

structure *n:* (geological) a formation of interest to drillers. For example, if a particular well is on the edge of a structure, the wellbore has penetrated the reservoir (structure) near its periphery.

stuck pipe *n:* drill pipe, drill collars, casing, or tubing having inadvertently become immovable in the hole. Sticking may occur when drilling is in progress, when casing is being run in the hole, or when the drill pipe is being hoisted.

stuck point *n:* the depth in the hole at which the drill stem, tubing, or casing is stuck.

stud-link chain *n:* (nautical) an anchor chain on which each link has a bar, or stud, across the shorter dimension of the link to prevent kinking and deformation under load.

stuffing box *n:* a packing gland screwed in the top of the wellhead through which the polished rod operates on a pumping well. It prevents the escape of oil, diverting it into a side outlet to which is connected the flow line, leading to the oil and gas separator or the field storage tank.

sub *n:* a short, threaded piece of pipe used to adapt parts of the drilling string that cannot otherwise be screwed together because of differences in thread size or design. A sub may also perform a special function. Lifting subs are used with drill collars to provide a shoulder to fit the drill pipe elevators. A kelly saver sub is placed between the drill pipe and the kelly to prevent excessive thread wear of the kelly and drill pipe threads. A bent sub is used when drilling a directional hole. Sub is a short expression for substitute.

sub elevator *n:* a small attachment on the rod-transfer equipment that picks up the rods after they are unscrewed from the string and then transfers them to the rod hanger, or reverses the procedure when going into the hole. See *rod-transfer equipment.*

submersible *n:* See *submersible drilling rig.*

submersible drilling rig *n:* an offshore drilling structure with several compartments that are flooded to cause the structure to submerge and rest on the seafloor. Most submersible rigs are used only in shallow waters.

submersible pump *n:* a pump that is placed below the level of fluid in a well. It is usually driven by an electric motor and consists of a series of rotating blades that impart centrifugal motion to lift the fluid to the surface.

subsea blowout preventer *n:* a blowout preventer placed on the seafloor for use by a floating offshore drilling rig.

subsea completion system *n:* a submersible apparatus similar to a bathysphere in which men are lowered to the ocean bottom to work on the wellheads of completed wells. The wellhead contains a cellar to which the system is attached, allowing the men to work in a dry atmosphere.

subsea test tree *n:* a device designed to be landed in a subsea wellhead or blowout preventer stack to provide a means of closing in the well on the ocean floor so that a drill stem test of an offshore well can be obtained.

substitute natural gas *n:* See *synthetic natural gas.*

substructure *n:* the foundation on which the derrick or mast and usually the drawworks sit; it contains space for storage and well control equipment.

subsurface *adj:* below the surface of the earth (e.g., sursurface rocks).

subsurface geology *n:* the study of rocks that lie beneath the surface of the earth.

subsurface safety valve n: See *tubing safety valve*.

subsurfacing sampling n: a procedure in which a bottomhole sampler is lowered into the well and filled with a sample that is representative of the reservoir conditions and that contains all the constituents of the fluid in their true proportions. Tests run on this sample help to obtain an accurate knowledge of the physical properties of the reservoir fluid under actual conditions.

sucker rod n: a special steel pumping rod. Several rods screwed together make up the mechanical link from the beam pumping unit on the surface to the sucker rod pump at the bottom of a well. Sucker rods are threaded on each end and manufactured to dimension standards and metal specifications set by the petroleum industry. Lengths are from 25 to 30 feet (about 8 m); diameter varies from ½ to 1⅛ inches (12-30 mm). There is also a continuous sucker rod (trade name: Corod).

sucker rod coupling n: an internally threaded fitting used to join sucker rods.

sucker rod pump n: the downhole assembly used to lift fluid to the surface by the reciprocating action of the sucker rod string. Basic components are barrel, plunger, valves, and hold-down. Two types of sucker rod pumps are the tubing pump, in which the barrel is attached to the tubing, and the rod, or insert, pump, which is run into the well as a complete unit.

sucker rod pumping n: a method of artificial lift in which a subsurface pump located at or near the bottom of the well and connected to a string of sucker rods is used to lift the well fluid to the surface. The weight of the rod string and fluid is counterbalanced by weights attached to a reciprocating beam or to the crank member of a beam pumping unit or by air pressure in a cylinder attached to the beam.

sucker rod whip n: an undesirable whipping motion in the sucker rod string that occurs when the string is not properly attached to the sucker rod pump or when the pump is operated at a resonant speed.

suction line n: the line that carries a product out of a tank to the suction side of the pumps; also called the loading line.

suction pit n: also called a suction tank, sump pit, or mud suction pit. See *suction tank*.

suction tank n: the mud tank from which mud is picked up by the suction of the mud pumps. Also called a suction pit.

suitcase sand n: a formation found to be nonproductive. When such a formation is encountered, operations are suspended, and the crews pack their suitcases and move to another job; hence, the name.

sulfamic acid n: a crystalline acid (NH_2SO_3H), a derivative of sulfuric acid that is sometimes used in acidizing.

sulfate-reducing bacteria n: bacteria that digest sulfate present in water, causing the release of hydrogen sulfide, which combines with iron to form iron sulfide, a troublesome scale.

sulfate resistance n: the ability of a cement to resist deterioration by sulfate ions.

sulfur n: a pale yellow, nonmetallic chemical element. In its elemental state, called free sulfur, it has a crystalline or amorphous form. In many gas streams, sulfur may be found as volatile sulfur compounds—hydrogen sulfide, sulfur oxides, mercaptans, carbonyl sulfide. Reduction of their concentration levels is necessary for corrosion control and, in many cases, necessary for health and safety reasons. Its symbol is S.

sulfur dioxide n: a colorless gaseous compound of sulfur and oxygen (SO_2) with the odor of rotten eggs. A product of the combustion of hydrogen sulfide, it is poisonous and irritating.

sulfuric acid n: a colorless, oily liquid compound of hydrogen, sulfur, and oxygen (H_2SO_4), strongly poisonous and corrosive. It is formed when hydrogen sulfide (H_2S) or sulfur dioxide (SO_2) is mixed with water (H_2O). Also called vitriolic acid.

sulfurous acid n: a colorless liquid compound of hydrogen, sulfur, and oxygen (H_2SO_3), weakly corrosive, with the odor of sulfur. It is formed when hydrogen sulfide (H_2S) or sulfur dioxide (SO_2) is mixed with water (H_2O).

sulfur plant n: a plant that makes sulfur from the hydrogen sulfide extracted from natural gas. One-third of the hydrogen sulfide is burned to sulfur dioxide, which reacts with the remaining hydrogen sulfide in the presence of a catalyst to make sulfur and water.

sul wtr abbr: sulfur water; used in drilling reports.

sump n: a low place in a vessel or tank, used to accumulate settlings that are later removed through an opening in the bottom of the vessel.

sump pit n: also called a suction pit. See *suction pit*.

supercharge v: to supply a charge of air to the intake of an internal-combustion engine at a pressure higher than that of the surrounding atmosphere.

supertanker *n:* a tanker with a capacity over 100,000 deadweight tons. Supertankers with a capacity larger than 100,000 dwt but less than 500,000 dwt are called *very large crude carriers*. Those with a capacity over 500,000 dwt are called *ultralarge crude carriers*.

sur *abbr:* survey; used in drilling reports.

surface-active agent *n:* See *surfactant*.

surface casing *n:* also called surface pipe. See *surface pipe*.

surface decompression *n:* a process used by a diver to eliminate inert gases from his tissues, whereby he breathes high partial pressures of oxygen while resting after a dive in order to reduce the risk of getting decompression sickness.

surface drilling unit *n:* an offshore drilling rig that is either a drill ship or a drilling barge; so called because the rig floats on the surface of the water.

surface-motion compensator *n:* a heave compensator.

surface pipe *n:* the first string of casing (after the conductor pipe) that is set in a well, varying in length from a few hundred to several thousand feet. Some states require a minimum length to protect freshwater sands. Compare *conductor pipe*.

surface pressure *n:* pressure measured at the wellhead.

surface readout device *n:* an electronic device in which a probe is inserted into the drill stem near a directional drilling deflection tool. The probe sends to the surface continuous signals that show the direction and angle at which the bit is drilling. Readout devices greatly simplify accurate orientation of the drilling assembly so that a number of directional surveys can be eliminated. See *directional drilling*.

surface safety valve *n:* a valve, mounted in the Christmas tree assembly, that stops the flow of fluids from the well if damage occurs to the assembly.

surface waste *n:* waste incurred by line leaks, seepage, inexpedient storage, and so forth. Usually such waste is regulated by federal or state agencies.

surfactant *n:* a substance that affects the properties of the surface of a liquid or solid by concentrating on the surface layer. Surfactants are useful in that their use can ensure that the surface of one substance or object is in thorough contact with the surface of another substance.

surfactant mud *n:* a drilling mud prepared by adding a surfactant to a water-base mud to change the colloidal state of the clay from that of complete dispersion to one of controlled flocculation. Such muds were originally designed for use in deep, high-temperature wells, but their many advantages (high chemical and thermal stability, minimum swelling effect on clay-bearing zones, lower plastic viscosity, etc.) extend their applicability.

surge *n:* 1. an accumulation of liquid above a normal or average level, or a sudden increase in its flow rate above a normal flow rate. 2. the motion of a mobile offshore drilling rig in a direction in line with the centerline of the rig, especially the front-to-back motion of the rig when it is moored in a seaway.

surge effect *n:* a rapid increase in pressure downhole that occurs when the drill stem is lowered rapidly or when the mud pump is quickly brought up to speed after starting.

surge tank *n:* a tank or vessel through which liquids or gases are passed to ensure steady flow and eliminate pressure surges.

surging *n:* a rapid increase in pressure downhole that occurs when the drill stem is lowered too fast or when the mud pump is brought up to speed after starting.

suspended S&W *n:* sediment and water that is suspended in oil and that can be separated only by (1) centrifuge with appropriate solvents or (2) extraction by distillation.

SW *abbr:* salt water; used in drilling reports.

swab *n:* a hollow, rubber-faced cylinder mounted on a hollow mandrel with a pin joint on the upper end to connect to the swab line. A check valve that opens upward on the lower end provides a way to remove the fluid from the well when pressure is insufficient to support flow. *v:* to operate a swab on a wireline to bring well fluids to the surface when the well does not flow naturally. Swabbing is a temporary operation to determine whether or not the well can be made to flow. If the well does not flow after being swabbed, a pump is installed as a permanent lifting device to bring the oil to the surface.

swabbed show *n:* formation fluid that is pulled into the wellbore because of an underbalance of formation pressure caused by pulling the drill string too fast.

swabbing effect *n:* a phenomenon characterized by formation fluids being pulled or swabbed into the wellbore when the drill stem and bit are pulled up the wellbore fast enough to reduce the hydrostatic pressure of the mud below the bit. If enough formation fluid is swabbed into the hole, a kick can result.

swabbing line n: also called sand line. See *sand line*.

swage n: a tool used to straighten damaged or collapsed casing in a well.

swage nipple n: a pipe fitting with external threads of different sizes on each end.

swamper n: (slang) a helper on a truck, tractor, or other machine.

sway n: the motion of a mobile offshore drilling rig in a linear direction from side to side or perpendicular to a line through the centerline of the rig; especially, the side-to-side motion when the rig is moored in a seaway.

swbd abbr: swabbed; used in drilling reports.

swbg abbr: swabbing; used in drilling reports.

sweet adj.: having an absence or near-absence of sulfur compounds, as defined by a given specification standard.

sweet corrosion n: the deterioration of metal caused by contact with carbon dioxide in water.

sweet crude n: also called sweet crude oil. See *sweet crude oil*.

sweet crude oil n: oil containing little or no sulfur, especially little or no hydrogen sulfide.

sweeten v: to remove sulfur or sulfur compounds from gas or oil.

sweet gas n: gas that has no more than the maximum sulfur content defined by (1) the specifications for the sales gas from a plant or (2) the definition by a legal body such as the Railroad Commission of Texas.

swelled box n: a box connection on a tool joint that has been belled by too much torque.

swingline n: an extension of the suction line that pivots vertically inside an oil tank. It reduces stratification by allowing an operator to withdraw product from varying heights in the tank. A swingline may be used in place of a mixing nozzle or other circulating system.

switch and control gear n: on a diesel-electric drilling rig, the equipment utilized to distribute and transmit electric power to the electric motors from the generators.

switcher n: (obsolete) lease operator or pumper. See *pumper*.

swivel n: a rotary tool that is hung from the rotary hook and traveling block to suspend and permit free rotation of the drill stem. It also provides a connection for the rotary hose and a passageway for the flow of drilling fluid into the drill stem.

swivel-connector grip n: a braided-wire device used to join the end of one wire rope to the end of another wire rope temporarily. When tension is put on this device, it stretches and grips the wire ropes firmly, allowing the wire rope to be threaded through the blocks. When tension is released, this device relaxes, allowing the rope to be released. Also called a snake or a swivel-type stringing grip.

swivel packing n: special rubberized compounds placed in a swivel to prevent drilling fluid under high pressure from leaking out.

swivel sub n: a sub containing a swivel joint, capable of permitting rotation between its two ends.

swivel-type stringing grip n: also called a swivel-connector grip. See *swivel-connector grip*.

sx abbr: sacks; used in drilling and mud reports.

synchronous adj: occurring at the same moment; for two or more generators, it means being phased together.

syncline n: a downward, trough-shaped configuration of folded, stratified rocks. Compare *anticline*.

synergistic effect n: the added effect produced by two processes working in combination, greater than the sum of the individual effects of each process.

synthetic natural gas n: a gas that is obtained either by heating coal or by refining heavier hydrocarbons. Hydrogen must be added to the product to make up for deficiencies in the original hydrocarbon source.

T *sym:* tesla; *abbr:* top of; used in drilling reports.

t *sym:* tonne.

TA *abbr:* temporarily abandoned.

tachometer *n:* an instrument that measures the speed of rotation; abbreviated as tach.

tag *v:* to touch an object downhole with the drill stem (as to tag the bottom of the hole or to tag the top of the fish).

tag line *n:* a utility rope or cable that is attached to unwieldy loads being hoisted by a crane to allow a load handler better control of the movement of the load.

tail chain *n:* a short length of chain that is attached to the end of a winch line, usually provided with a special hook that fastens to objects.

tail gas *n:* the exit gas from a plant.

tail gate *n:* the point in a gas processing plant at which the residue gas is last metered, usually the plant residue sales meter or the allocation meter.

tail out rods *v:* to pull the bottom end of a sucker rod away from a well when laying rods down.

tail pipe *n:* 1. a pipe run in a well below a packer. 2. a pipe used to exhaust gases from the muffler of an engine to the outside atmosphere.

tail roller *n:* a larger roller located across the stern of an anchor-handling boat, over which pendant lines travel when an anchor is being brought in or dropped.

take-or-pay clause *n:* a contract clause that guarantees pay to a seller for gas, even though the particular gas volume is not taken during a specified time period. Some contracts stipulate a time period for the buyer to take later delivery of the gas without penalty.

take out *v:* to remove a joint or stand of pipe from the drill stem.

tally *v:* to measure and record the total length of pipe, casing, or tubing that is to be run in a well.

tang *n:* a piece that provides an extension of an instrument (such as a file) and serves to form the handle or make a connection for the attachment of a handle.

tank *n:* a metal, plastic, or wooden container used to store a liquid. Three types include mud tanks for drilling, production tanks, and storage tanks.

tankage *n:* the total capacity of a number of tanks in a field.

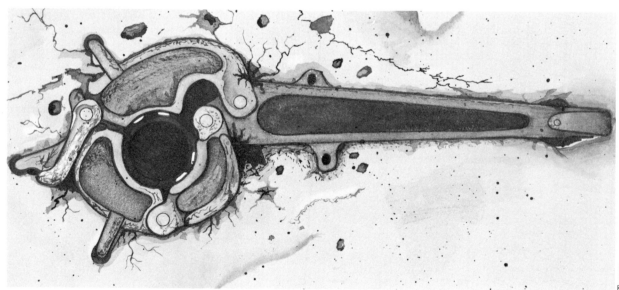
Tongs

tank barge *n:* a large, flat-bottomed vessel divided into compartments and used to carry crude or fuel oil.

tank battery *n:* a group of production tanks located in the field to store crude oil.

tank bottoms *n pl:* the settlings in the bottom of a storage tank. See *basic sediment and water* and *bottoms*.

tank car *n:* a railroad car used to transport petroleum or petroleum products.

tanker *n:* a ship designed to transport oil, LPG, LNG, or SNG; also called a tank ship. Tankers whose capacity is 100,000 deadweight tons or more are supertankers, either very large crude carriers or ultralarge crude carriers.

tank farm *n:* a group of large tanks maintained by a pipeline and used to store oil after it has been transferred from the production tanks and before it is transported to the refinery.

tank ship *n:* See *tanker*.

tank strapper *n:* the person who measures a tank at various levels to see how much it will hold.

tank table *n:* a table giving the number of barrels of fluid contained in a storage tank corresponding to the linear measurement on a gauge line. Tank tables are prepared from tank strapping measurements. See *strap*.

tank truck *n:* a truck designed to transport petroleum or petroleum products.

tap *n:* 1. a tool for forming an internal screw thread, consisting of a hardened tool-steel male screw grooved longitudinally so as to have cutting edges. 2. a hole or opening in a line or vessel into which a gauge or valve may be inserted and screwed tight.

tapered string *n:* drill pipe, tubing, sucker rods, and so forth with a diameter near the top of the well larger than the diameter below.

taper tap *n:* a tap having a gradually decreasing diameter from the top and used as a fishing tool for a hollow fish (such as a drill collar). The taper tap is run into the hollow fish and rotated to cut enough threads to provide a firm grip, permitting the fish to be pulled and recovered. See *tap*.

tape wrapping *n:* rolls of plastic sheeting with a preapplied adhesive, used to coat buried pipelines in order to prevent corrosion.

tariff *n:* the rate set by pipeline companies for moving oil.

tar sand *n:* a sandstone that chiefly contains very heavy, tarlike hydrocarbons. Tar sands are difficult to produce by ordinary methods; thus it is costly to obtain usable hydrocarbons from them.

taut-line position-reference system *n:* a system for monitoring the position of a floating offshore drilling rig in relation to the subsea wellhead by stretching a taut steel line from the rig to the ocean floor. An inclinometer measures the slope of the line at the rig, and, because the line is assumed to have been straight when stretched from the rig to the ocean floor, any angle in the line indicates that the rig has moved. The system's weakness is that the taut line can be distorted by currents and thus give inaccurate readings. Compare *acoustic position reference* and *position-reference system*.

Tcf *abbr:* trillion cubic feet.

Tcf/d *abbr:* trillion cubic feet per day.

TD *abbr:* total depth.

TDC *abbr:* top dead center.

tear down *v:* also called rig down. See *rig down*.

TEG *abbr:* triethylene glycol.

telemetry *n:* the process of gathering data by electronic or other kinds of sensing devices and transmitting that data to remote points.

telescoping derrick *n:* a portable mast that is capable of being erected as a unit, usually by a tackle that hoists the wireline or by hydraulic pistons. Generally the upper section of a telescoping derrick is nested (telescoped) inside the lower section of the structure and raised to full height either by the wireline or by a hydraulic system.

telescoping joint *n:* a device used in the marine riser system of a mobile offshore drilling rig to compensate for the vertical motion of the rig caused by wind, waves, or weather. It consists of an inner barrel attached beneath the rig floor and an outer barrel attached to the riser pipe and is an integrated part of the riser system.

telltale hole *n:* a hole drilled into the space between rings of packing material used with a liner in a mud pump. When the liner packing fails, fluid spurts out of the telltale hole with each stroke of the piston, indicating that the packing must be renewed.

temperature *n:* a measure of heat or the absence of heat, expressed in degrees Fahrenheit or Celsius. The latter is the standard used in countries on the metric system.

temperature correction factor n: a factor for correcting volumes of gas to the volume occupied at a specific reference temperature. Reference temperature most commonly used in the petroleum industry is 60°F (15.56°C).

temperature gradient n: 1. the rate of change of temperature with displacement in a given direction. 2. the increase in temperature of a well as its depth increases.

temperature log n: a survey run in cased holes to locate the top of the cement in the annulus. Since cement generates a considerable amount of heat when setting, a temperature increase will be found at the level where cement is found behind the casing.

temperature survey n: an operation used to determine temperatures at various depths in the wellbore. In addition, it is used to determine the height of cement behind the casing and to locate the source of water influx into the wellbore.

temper (or temple) screw n: a part on a cable-tool rig used to regulate the force of the blow delivered to the drill bit. Attached to the walking beam, it controls the feed rate of the drilling tools.

template n: See *temporary guide base*.

temporarily abandoned adj: temporarily shut in but not plugged.

temporary guide base n: the initial piece of equipment lowered to the ocean floor once a mobile offshore drilling rig has been positioned on location. It serves as an anchor for the guidelines and as a foundation for the permanent guide base and has an opening in the center through which the bit passes. It is also called a template.

tender n: 1. the barge anchored alongside a relatively small offshore drilling platform, usually containing living quarters, storage space, and the mud system. 2. a shipment of oil presented by a shipper to a pipeline for movement. 3. a form required by certain regulatory bodies in some states for their approval of products shipped from plants or other sources. 4. the person responsible for tending to a diver's needs.

tensile adj: of or relating to tension.

tensile strength n: the greatest longitudinal stress that a metal can bear without tearing apart. Tensile strength of a metal is greater than yield strength.

tensile stress n: stress developed by a material bearing a tensile load. See *stress*.

tension n: the condition of a string, wire, pipe, or rod that is stretched between two points.

tensioner system n: a system of devices installed on a floating offshore drilling rig to maintain a constant tension on the riser pipe despite any vertical motion made by the rig. The guidelines must also be tensioned, and a separate tensioner system is provided for them.

terminal n: a point to which oil is transported through pipelines. It usually includes a tank farm and may include tanker-loading facilities.

tertiary recovery n: 1. the use of advanced enhanced oil recovery methods that not only restore formation pressure but also improve displacement of oil by overcoming forces that keep the oil trapped in rock pores. 2. the use of any enhanced oil recovery method to remove additional oil after secondary recovery. See *secondary recovery* and *enhanced oil recovery*.

tesla n: the unit of magnetic-flux density in the metric system. Its symbol is T.

test separator n: an oil and gas separator that is used to separate relatively small quantities of oil and gas, which are diverted through the testing devices on a lease.

test well n: a wildcat well.

tethered diving n: diving in which an umbilical hose is used to connect a diver to his gas supply.

tetraethyl lead n: an antiknock compound.

Texas deck n: the main load-bearing deck of an offshore drilling structure and the highest above the water, excluding auxiliary decks such as the helicopter landing pad.

theoretical gallons n pl: content of liquefiable hydrocarbons in a volume of gas, determined from analyses or tests of the gas.

therm n: a unit of gross heating value equal to 100,000 Btu ($1.055056 \cdot 10^8$ J).

thermal cracking n: the process of making oils of low boiling range (100°F to 550°F), to be used for motor fuels and burning oils, from oils of high boiling range (550°F to 800°F), such as gas oil and fuel oil. All modern commercial methods get this breaking-down action by subjecting the high boiling oils to high temperatures. Pressure up to 1,000 psi is used for producing a dense system and making sure of good contact within the desired temperature and in an apparatus of reasonable size.

Thermal Decay Time Log n: proprietary name for a type of pulsed-neutron survey.

thermalization n: the process of reducing the energy of neutrons to thermal energy levels.

thermal neutron n: neutron having an average energy level at room temperature of 0.025 ev.

thermal neutron population n: the number of thermal neutrons around the neutron logging tool. The number of capture gamma rays present at any time is directly proportional to the thermal neutron population.

thermal recovery n: a method of enhanced oil recovery in which heat is introduced into a reservoir to lower the viscosity of heavy oils and facilitate their flow into producing wells. The pay zone may be heated by injecting steam (steam drive) or by injecting air and burning a portion of the oil in place (in situ combustion). See *steam drive* and *in situ combustion*.

thermocouple n: a device consisting of two dissimilar metals bonded together, with electrical connections to each. When the device is exposed to heat, an electrical current is generated, the magnitude of which varies with the temperature. It is used to measure temperatures higher than those that can be measured by an ordinary thermometer, such as those in an engine exhaust.

thermometer n: an instrument that measures temperature. Thermometers provide a way to estimate temperature from its effect on a substance with known characteristics (such as a gas that expands when heated). Various types of thermometers measure temperature by measuring the change in pressure of a gas kept at a constant volume, the change in electrical resistance of metals, or the galvanic effect of dissimilar metals in contact. The most common thermometer is the mercury-filled glass tube that indicates temperature by the expansion of the liquid mercury.

thermoplastics n pl: a variety of materials often used in pipe coatings, whose molecular structure allows them to repeatedly soften when heated and harden when cooled.

thermosetting plastics n pl: plastics that solidify when first heated under pressure, but whose original characteristics are destroyed when remelted or remolded.

thermostat n: a control device used to regulate temperature.

thermowell n: a well in a process vessel or line used as a thermometer or thermocouple holder.

thickening time n: the amount of time required for cement to reach an API-established degree of consistency, or thickness. Thickening time begins when the slurry is actually mixed.

thief n: a device that is lowered into a tank to take an oil sample at any desired depth. The sample will subsequently be used to determine the BS&W content of the oil in the tank.

thief formation n: a formation that absorbs drilling fluid as the fluid is circulated in the well; also called a thief sand or a thief zone. Lost circulation is caused by a thief formation.

thief sand n: See *thief formation*.

thin v: to add a substance such as water or a chemical to drilling mud to reduce its viscosity.

thinning agent n: a special chemical or combination of chemicals that, when added to a drilling mud, reduces its viscosity.

thixotropy n: the property exhibited by a fluid that is in a liquid state when flowing and in a semisolid, gelled state when at rest. Most drilling fluids must be thixotropic so that the cuttings in the fluid will remain in suspension when circulation is stopped.

thread n: a continuous helical rib, as on a screw or pipe.

thread profile gauge n: a device to measure the amount of wear or stretch on pipe threads.

thread protector n: a device that is screwed onto or into pipe threads to protect the threads from damage when the pipe is not in use. Protectors may be metal or plastic.

thribble n: a stand of pipe made up of three joints and handled as a unit. Compare *single*, *double*, and *fourble*.

thribble board n: the name used for the working platform of the derrickman, or monkeyboard, when it is located at a height in the derrick equal to three lengths of pipe joined together.

throttling n: the choking or failing that occurs when a mud pump fails to deliver a full amount of fluid through one or more of its valves. Throttling is usually caused by improper lift of the valve.

throw n: the distance from the centerline of the main bearing of a crankshaft to the centerline of the connecting-rod journal. Two times the throw equals the stroke.

throw the chain n: to flip the spinning chain up from a tool joint box so that the chain wraps around the tool joint pin after it is stabbed into the box. The stand or joint of drill pipe is turned or spun by a pull on the spinning chain from the cathead on the drawworks.

thrust n: the force that acts on a shaft longitudinally.

thruster *n*: See *dynamic positioning*.

thumper *n*: a hydraulically operated hammer used in obtaining a seismograph in oil exploration. It is mounted on a vehicle and, when dropped, creates shock waves in subsurface formations, which are recorded and interpreted to reveal geological information.

tie-back string *n*: casing that is run from the top of a liner to the surface. A tie-back string is often used to provide a production casing that has not been drilled through.

tie-down *n*: a device to which a guy wire or brace may be attached, such as the anchoring device for the deadline of a hoisting-block arrangement.

tight formation *n*: a petroleum- or water-bearing formation of relatively low porosity and permeability.

tight hole *n*: 1. a well about which information is restricted for security or competitive reasons and such information given only to those authorized to receive it. 2. a section of the hole that, for some reason, is undergauge. For example, a bit that is worn undergauge will drill a tight hole.

tight spot *n*: a section of a borehole in which excessive wall cake has built up, reducing the hole diameter and making it difficult to run the tools in and out. Compare *key seat*.

time release *n*: feature built into oil field inhibitors that allows them to be introduced into production systems and their active ingredients released at certain timed intervals.

timing *n*: the relationship of all moving parts in an engine. Each part depends on another, so all parts must operate in the right relation with each other as the engine turns.

tin hat *n*: See *hard hat*.

tolerance *n*: the range of variation permitted in maintaining a specific dimension in machining a part. For example, a shaft measurement could be 2.000 in. ± 0.001 (50 mm ± 0.025). This means that any measurement from 1.999 in. to 2.001 in. (49.925 mm to 50.025 mm) would be acceptable.

ton *n*: 1. (nautical) a volume measure equal to 100 ft³ applied to mobile offshore drilling rigs. 2. (metric) a measure of weight equal to 1 000 kg. Usually spelled *tonne*.

tong dies *n*: very hard and brittle pieces of serrated steel that are installed in the tongs and that grip or bite into the tool joint of drill pipe when the tongs are latched onto the pipe.

tongman *n*: the member of the drilling crew who handles the tongs.

tongs *n pl*: the large wrenches used for turning in, making up, or breaking out drill pipe, casing, tubing, or other pipe; variously called casing tongs, pipe tongs, and so forth, according to the specific use. Power tongs are pneumatically or hydraulically operated tools that serve to spin the pipe up tight and, in some instances, to apply the final makeup torque.

ton-mile *n*: the unit of service given by a hoisting line in moving 1 ton of load over a distance of 1 mile.

tonnage *n*: (nautical) the size of a ship or spaces within a ship as measured in tons.

tonne *n*: a mass unit in the metric system equal to 1 000 kg.

tool dresser *n*: a driller's helper on a cable-tool rig, once responsible for sharpening or dressing the drill bit; sometimes called a toolie.

toolhouse *n*: a building for storing tools.

toolie *n*: (slang) tool dresser.

tool joint *n*: a heavy coupling element for drill pipe, made of special alloy steel. Tool joints have coarse, tapered threads and seating shoulders designed to sustain the weight of the drill stem, withstand the strain of frequent coupling and uncoupling, and provide a leakproof seal. The male section of the joint, or the pin, is attached to one end of a length of drill pipe, and the female section, or box, is attached to the other end. The tool joint may be welded to the end of the pipe, screwed on, or both. A hard-metal facing is often applied in a band around the outside of the tool joint to enable it to resist abrasion from the walls of the borehole.

toolpusher *n*: an employee of a drilling contractor who is in charge of the entire drilling crew and the drilling rig. Also called a drilling foreman, rig manager, rig supervisor, or rig superintendent.

top dead center *n*: the position of a piston when it is at the highest point possible in the cylinder of an engine, often marked on the flywheel.

top hold-down *n*: a mechanism for anchoring a sucker rod pump to the tubing, located at the top of the working barrel. Compare *bottom hold-down*.

top wiper plug *n*: a device placed in the cementing head and run down the casing behind cement to clean the cement off the walls of the casing and to prevent contamination between the cement and the displacement fluid.

torque *n*: the turning force that is applied to a shaft or other rotary mechanism to cause it to rotate or tend to do so. Torque is measured in units of length and force (foot-pounds, newton-metres).

torque converter *n:* a hydraulic device connected between an engine and a mechanical load such as a compound. Torque converters are characterized by an ability to increase output torque as the load causes a reduction in speed. Torque converters are used on mechanical rigs that have compounds.

torque indicator *n:* an instrument that measures the amount of torque (turning or twisting action) applied to the drill or casing string. The amount of torque applied to the string is important when joints are being made up.

torque recorder *n:* an instrument that measures and makes a record of the amount of torque (turning or twisting action) applied to the drill or casing string.

torsion *n:* twisting deformation of a solid body about an axis in which lines that were initially parallel to the axis become helices. Torsion is produced when part of the pipe turns or twists in one direction while the other part remains stationary or twists in the other direction.

total calculated volume *n:* the total volume of all petroleum liquids and sediment and water, corrected by the appropriate temperature correction (C_{tl}) for the observed temperature and API gravity, relative density, or density to a standard temperature such as 60°F or 15°C, and also corrected by the applicable pressure factor (C_{pl}) and meter factor and all free water measured at observed temperature and pressure. Total calculated volume is equal to the gross standard volume plus free water volume.

total depth *n:* the maximum depth reached in a well.

total observed volume *n:* the total measured volume of all petroleum liquids, sediment and water, and free water at observed temperature and pressure.

total S&W *n:* all sediment and water in a vessel's cargo tanks, whether settled or suspended.

tour *n:* (pronounced "tower") a working shift for drilling crew or other oil field workers. The most common tour is 8 hours long; the three daily tours are called daylight, evening, and graveyard (or morning). Sometimes 12-hour tours are used, especially on offshore rigs; they are called simply day tour and night tour.

tower *n:* 1. a vertical vessel such as an absorber, fractionator, or still. 2. a cooling tower.

TP *abbr:* tubing pressure; used in drilling reports.

tracer *n:* a substance added to reservoir fluids to permit the movements of the fluid to be followed or traced. Dyes and radioactive substances are used as tracers in underground water flows and sometimes helium is used in gas. When samples of the water or gas taken some distance from the point of injection reveal signs of the tracer, the route of the fluids can be mapped.

tracer log *n:* a survey that uses a radioactive tracer such as a gas, liquid, or solid having a high gamma ray emission. When the material is injected into any portion of the wellbore, the point of placement or movement can be recorded by a gamma ray instrument. The tracer log is used to determine channeling or the travel of squeezed cement behind a section of perforated casing. Also called tracer survey.

tracer survey *n:* also called tracer log. See *tracer log*.

tracking *n:* a rare type of bit-tooth wear occurring when the pattern made on the bottom of the hole by all three cones of a rock bit matches the bit-tooth pattern to such an extent that the bit gears itself to the formation and drills very little.

traction motor *n:* a direct current electric motor nominally found in railway use, but frequently adapted to power certain drilling rigs.

trammel *n:* a metal rod of precise length used to measure distance between two points where accessibility is limited; abbreviated as tram. It is often used to mark crankshaft positions on engines.

transducer *n:* a device actuated by power from one system and supplying power to another system, usually in a different form. For example, a telephone receiver receives electric power and supplies acoustic power.

transition zone *n:* 1. the area in which underground pressures begin to change from normal to abnormally high as a well is being deepened. 2. the areas in the drill stem near the point where drill pipe is made up on drill collars.

transmission *n:* the gear or chain arrangement by which power is transmitted from the prime mover to the drawworks, mud pump, or rotary table of a drilling rig.

transmission line *n:* 1. a high-voltage line used to transmit electric power from one place to another. 2. a pipeline used to transmit natural gas or other fluids.

trap *n:* layers of buried rock strata that are arranged so that petroleum accumulates in them.

traveling barrel pump *n:* a sucker rod insert pump in which the working barrel travels and the plunger remains stationary. The working barrel is connected to the sucker rod string through a connector and the traveling valve; the standing valve is connected to the top of the plunger, which in turn is connected to the bottom hold-down.

traveling block *n:* an arrangement of pulleys, or sheaves, through which drilling line is reeved and which moves up and down in the derrick or mast. See *block*.

traveling valve *n:* one of the two valves in a sucker rod pumping system. The traveling valve moves with the movement of the sucker rod string. On the upstroke, the ball member of the valve is seated, supporting the fluid load. On the downstroke, the ball is unseated, allowing fluid to enter into the production column. Compare *standing valve*.

tray *n:* a horizontal device in a tower that holds liquid and provides a vapor-liquid contact step. Several common types are bubble-cap, perforated, or valve trays.

trayed column *n:* a vessel wherein gas and liquid, or two miscible liquids, are contacted on trays, usually countercurrently.

tread diameter *n:* the diameter of the sheave of a drilling block, measured from groove bottom to groove bottom. It is the diameter of the circle around which the wire rope is bent.

treat *v:* to subject a substance to a process or to a chemical reagent to improve its quality or remove a contaminant.

treater *n:* a vessel in which oil is treated for the removal of BS&W or other objectionable substances by the addition of chemicals, heat, electricity, or all three.

tricone bit *n:* a type of bit in which three cone-shaped cutting devices are mounted in such a way that they intermesh and rotate together as the bit drills. The bit body may be fitted with nozzles, or jets, through which the drilling fluid is discharged. A one-eyed bit is used in soft formations to drill a deviated hole.

triethylene glycol *n:* a liquid chemical used in gas processing to remove water from the gas. See *glycol dehydration*.

trim *n:* the difference between fore and aft draft readings on a marine vessel or an offshore drilling rig.

trim correction *n:* a correction applied to cargo ullages when the vessel is not on an even keel. Valid only when liquid completely covers tank bottom and is not in contact with the underside of the deck.

trip *n:* the operation of hoisting the drill stem from and returning it to the wellbore. *v:* shortened form of "make a trip". See *make a trip*.

trip gas *n:* an accumulation of gas, usually a negligible amount, that enters the hole when a trip is being made.

trip in *v:* See *go in the hole*.

triplex pump *n:* a reciprocating pump with three pistons or plungers.

trip out *v:* See *come out of the hole*.

tripping *n:* the operation of hoisting the drill stem out of and returning it to the wellbore; making a trip. See *make a trip*.

trip tank *n:* a small mud tank with a capacity of 10 to 15 bbl, usually with 1-bbl divisions, used exclusively to ascertain the amount of mud necessary to keep the wellbore full with the exact amount of mud that is displaced by drill pipe. When the bit comes out of the hole, a volume of mud equal to that which the drill pipe occupied while in the hole must be pumped into the hole to replace the pipe. When the bit goes back in the hole, the drill pipe displaces a certain amount of mud, and a trip tank again can be used to keep track of this volume.

tritium *n:* an isotope of hydrogen with two neutrons in the nucleus designated $_1H^3$.

truck-mounted rig *n:* a well-servicing and workover rig that is mounted on a truck chassis.

true-to-gauge hole *n:* a hole that is the same size as the bit that was used to drill the hole; frequently referred to as a full-gauge hole.

true vertical depth *n:* the depth of a well measured from the surface straight down to the bottom of the well. The true vertical depth of a well may be quite different from its actual measured depth, because wells are very seldom drilled exactly vertical.

trunk line *n:* a main line.

tstg *abbr:* testing; used in drilling reports.

tube *v:* to run tubing in a well.

tube bundle *n:* the inner piping of a condenser or heat exchanger, typically consisting of a group of pipes placed inside the shell of a tank.

tube sheet *n:* a metal plate through which the tubes in the tube bundles are placed for support, effecting a pressure-tight connection between the tubes and the heads of a condenser or heat exchanger.

tubing *n:* small-diameter pipe that is run into a well to serve as a conduit for the passage of oil and gas to the surface.

tubing elevators *n pl:* a clamping apparatus used to pull tubing. The elevators latch onto the pipe just below the top collar. The elevators are attached by steel links or bails to the hook.

tubing hanger *n:* an arrangement of slips, built into a steel housing and engaged in the upper end of the wellhead, that serves as a support for the suspended tubing string.

tubing head *n:* a flanged fitting that supports the tubing string, seals off pressure between the casing and the outside of the tubing, and provides a connection that supports the Christmas tree.

tubing job *n:* the act of pulling tubing out of and running it back into a well.

tubingless completion *n:* a method of producing a well in which only a small-diameter production casing is set through the pay zone, with no tubing or inner production string used to bring formation fluids to the surface. This type of completion has limited application in small-volume, dry-gas reservoirs.

tubing pressure *n:* pressure on the tubing in a well at the wellhead.

tubing pump *n:* a sucker rod pump in which the barrel is attached to the tubing. See *sucker rod pump.*

tubing safety valve *n:* a device installed in the tubing string of a producing well to shut in the flow of production if the flow exceeds a preset rate. Tubing safety valves are widely used in offshore wells to prevent pollution if the wellhead fails for any reason.

tubing slips *n:* slips designed specifically to be used with tubing.

tubing spider *n:* a device used with slips to prevent tubing from falling into the hole when a joint of pipe is being unscrewed and racked.

tubing tongs *n pl:* large wrenches used to break out and make up tubing. They may be operated manually, hydraulically, or pneumatically.

tubular goods *n pl:* any kind of pipe; also called tubulars. Oil field tubular goods include tubing, casing, drill pipe, and line pipe.

tubulars *n pl:* shortened form of tubular goods.

tungsten carbide *n:* a fine, very hard gray crystalline powder, a compound of tungsten and carbon. This compound is bonded with cobalt or nickel in cemented carbide compositions and used for cutting tools, abrasives, and dies.

tungsten carbide bit *n:* a type of roller cone bit with inserts made of tungsten carbide. Also called tungsten carbide insert bit.

tungsten carbide insert bit *n:* also called tungsten carbide bit. See *tungsten carbide bit.*

turbine motor *n:* usually called a turbodrill. See *turbodrill.*

turbocharger *n:* a centrifugal blower driven by exhaust gas turbines and used to supercharge an engine.

turbodrill *n:* a drilling tool that rotates a bit that is attached to it by the action of drilling mud on the turbine blades built into the tool. When a turbodrill is used, rotary motion is imparted only at the bit; therefore, it is unnecessary to rotate the drill stem. Although straight holes can be drilled with the tool, it is used most often in directional drilling.

turboexpander *n:* a device that converts energy of a gas or vapor stream into mechanical work by expanding the gas or vapor through a turbine.

turbulent flow *n:* the flow of a fluid in an erratic, nonlinear motion, caused by high velocity.

turnaround *n:* 1. space that permits the turning around of vehicles, specifically on the drill site. 2. a period during which a plant is shut down completely for repairs, inspections, or modifications that cannot be made while the plant is operating.

turnkey contract *n:* a drilling contract that calls for the payment of a stipulated amount to the drilling contractor on completion of the well. In a turnkey contract, the contractor furnishes all material and labor and controls the entire drilling operation, independent of supervision by the operator.

turn to the right *v:* on a rotary rig, to rotate the drill stem clockwise. When drilling ahead, the expression "on bottom and turning to the right" indicates that drilling is proceeding normally.

turret mooring *n:* a system of mooring a drill ship on the drilling site, in which mooring lines are spooled onto winches mounted on a turret in the center of the vessel. Because all mooring lines are connected to the turret, the vessel is free to rotate around the turret axis and head into oncoming seas, regardless of direction.

TVD *abbr:* true vertical depth.

twin *n:* a well drilled on the same location as another well and closely offsetting it, but producing from a different zone.

twistoff *n:* a complete break in pipe caused by rotational force wrenching damaged pipe apart.

twist off *v:* to part or split drill pipe or drill collars, primarily because of metal fatigue in the pipe or because of mishandling.

two-phase flow *n:* a flow of a vapor phase and a liquid phase in the same pipeline.

two-step grooving system *n:* a pattern of drum spooling in which the wire rope is controlled by grooves to move parallel to drum flanges for half the circumference (180 degrees) and then crosses over to start the next wrap. Also called the counterbalance system.

two-stroke/cycle engine *n:* an engine in which the piston moves from top dead center to bottom dead center and then back to top dead center to complete a cycle. Thus, the crankshaft must turn one revolution, or 360 degrees.

Twp *abbr:* township; used in drilling reports.

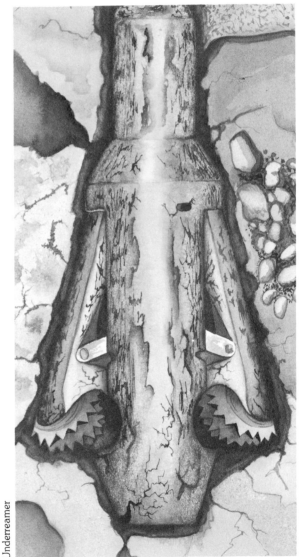
Underreamer

ULCC *abbr:* ultralarge crude carrier.

ullage *n:* the amount by which a tank or a vessel comes short of being full, especially on ships. Ullage in a tank is necessary to allow space for the expansion of the oil in the tank when the temperature increases. Also called outage.

ultimate recovery *n:* total anticipated recovery of oil or gas from a well, lease, or pool.

ultralarge crude carrier *n:* a supertanker whose capacity is 500,000 deadweight tons or more. Compare *very large crude carrier*.

ultrasonic devices *n:* corrosion-monitoring devices that transmit ultrasonic waves through production structures in order to locate discontinuities in metal structure, which indicate corrosion damage.

umbilical *n:* a line that supplies a diver or a diving bell with a lifeline, a breathing gas, communications, a pneumofathometer, and, if needed, a heat supply.

unconf *abbr:* unconformity; used in drilling reports.

unconformity *n:* 1. lack of continuity in deposition between rock strata in contact with one another, corresponding to a gap in the stratigraphic record. 2. the surface of contact between rock beds in which there is a discontinuity in the ages of the rocks. See *angular unconformity* and *disconformity*.

unconsolidated sandstone *n:* a sand formation in which individual grains do not adhere to one another. If an unconsolidated sandstone produces oil or gas, it will produce sand as well if not controlled or corrected.

underflow *n:* the lower discharge stream in a cone-shaped centrifuge, moving downward and leaving by way of the apex.

undergauge bit *n:* a bit whose outside diameter is worn to the point at which it is smaller than it was when new. A hole drilled with an undergauge bit is said to be undergauge.

underground blowout *n:* an uncontrolled flow of gas, salt water, or other fluid out of the wellbore and into another formation that the wellbore has penetrated.

underground waste *n:* recoverable reserves lost as a result of damage to the reservoir.

underream *v:* to enlarge the wellbore below the casing.

underreamer *n:* a device used to underream. See *underream*.

union *n:* a coupling device that allows pipes to be connected without being rotated. The mating surfaces are pulled together by a flanged, threaded collar on the union.

unit *n:* 1. a piece or several pieces of equipment performing a complete function such as a beam pumping unit. 2. several leases that are operated by one company. 3. one lease that is operated by several companies.

United States Geological Survey *n:* a governmental agency responsible for the enforcement of rules pertaining to the drilling and production of oil and gas in offshore areas of the United States.

unit operator *n:* the oil company in charge of development and production in an oil field in which several companies have joined together to produce the field.

unproven area *n:* a wildcat area.

unsaturated hydrocarbon *n:* a straight-chain compound of hydrogen and carbon whose total combining power has not yet been reached and to which other atoms or radicals can be added.

upper kelly cock *n:* the kelly cock, as distinguished from the drill stem safety valve, sometimes called the lower kelly cock. See *kelly cock*.

upper tier *n:* a category of oil production for purposes of price control. Upper tier refers to oil that comes from reservoirs that began producing subsequent to 1972. It also refers to oil produced from wells having a production rate of 10 bbl/day or less for a continuous period of 12 months. Also called stripper oil.

upset *v:* to forge the ends of tubular products so that the pipe wall acquires extra thickness and strength near the end. Usually upsetting is performed to provide the thickness needed to form threads so that the tubular goods can be connected. *n:* the thickened area formed by upsetting of tubular goods.

upstream *adv:* in the direction opposite the flow in a line. *n:* the point in a line or system situated opposite the direction of flow.

urea *n:* a soluble, weakly basic, nitrogenous compound, $CO(NH_2)_2$, that is used in the manufacture of resins and plastics.

use tests *n:* periodic equipment inspections to determine corrosion damage in fields that have been producing for several years.

USGS *abbr:* United States Geological Survey.

U-tube *n:* a U-shaped tube.

U-tubing *n:* the action of fluids flowing in a U-tube (as heavy mud forcing lighter mud down the drill stem and up the annulus).

V *sym:* volt.

vacuum *n:* 1. theoretically, a space that is devoid of all matter and that exerts zero pressure. 2. a condition that exists in a system when pressure is reduced below atmospheric pressure.

vacuum degasser *n:* a device in which gas-cut mud is degassed by the action of a vacuum inside a tank. The gas-cut mud is pulled into the tank, the gas removed, and the gas-free mud discharged back into the mud tank.

vacuum gauge *n:* an instrument used on gas or gasoline engines to indicate the performance characteristics and load.

valve *n:* a device used to control the rate of flow in a line, to open or shut off a line completely, or to serve as an automatic or semiautomatic safety device. Those with extensive usage include the gate valve, plug valve, globe valve, needle valve, check valve, and pressure relief valve.

valve tray *n:* a tray installed in an absorber tower or fractionating column, similar to a bubble cap tray except that the passageways permitting gas flow upward through the tray possess valves that reduce the size of the passages when the flow rate is reduced. Valve trays are more common than bubble cap trays because they are more efficient and less expensive.

vapor *n:* a substance in the gaseous state, capable of being liquefied by compression or cooling.

vaporization *n:* 1. the act or process of converting a substance into the vapor phase. 2. the state of substances in the vapor phase.

vapor-liquid equilibrium ratio *n:* the partition coefficient, usually designated as K, which is equivalent to y/xm where y is the mol fraction of a given component in the vapor phase that is in equilibrium with x, the mol fraction of the same component in the liquid phase. K is a function of temperature, pressure, and composition of the particular system.

vapor phase *n:* the existence of a substance in the gaseous state.

vapor pressure *n:* the pressure exerted by the vapor of a substance when the substance and its vapor are in equilibrium. Equilibrium is established when the rate of evaporation of a substance is equal to the rate of condensation of its vapor.

vaporproof *adj:* not susceptible to or affected by vapors. For example, an electrical switch is made vaporproof so that a spark issuing from it will not cause an explosion in the presence of combustible gases.

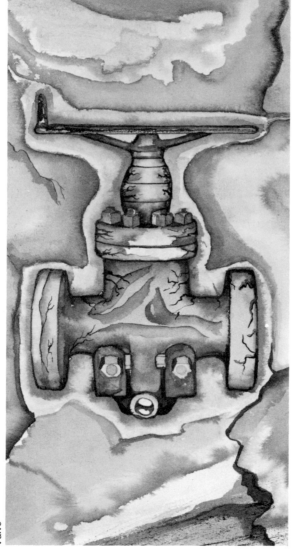

Valve

vapor recovery *n:* a system or method by which vapors are retained and conserved.

vapor recovery unit *n:* in petroleum refining, a process unit consisting of a scrubber and a compressor, designed to recover petroleum from hydrocarbon vapors and to handle safely toxic gases produced from some wells.

variation *n:* the angle by which a compass needle deviates from true north.

V-belt *n:* a belt with a trapezoidal cross section, made to run in sheaves, or pulleys, with grooves of corresponding shape.

V-door *n:* an opening at floor level in a side of a derrick or mast. The V-door is opposite the drawworks and is used as an entry to bring in drill pipe, casing, and other tools from the pipe rack. The name comes from the fact that on the old standard derrick, the shape of the opening was an inverted V.

vent *n:* an opening in a vessel, line, or pump to permit the escape of air or gas.

venturi effect *n:* the drop in pressure resulting from the increased velocity of a fluid as it flows through a constricted section of a pipe.

venturi tube *n:* a short tube with a calibrated constriction in it that is used in instruments or devices such as jet hoppers; developed according to the principle that a fluid flowing through a constriction has increased velocity and reduced pressure.

venturi-tube meter *n:* a flowmeter used to determine the rate of flow and employing a venturi tube as the primary element for creating differential pressures in flowing gases or liquids. Compare *orifice meter*.

vertical *n:* an imaginary line at right angles to the plane of the horizon. *adj:* of a wellbore, straight, not deviated.

vertically integrated oil company *n:* a large oil company that is involved in all aspects of the oil industry.

very large crude carrier *n:* a supertanker whose capacity is larger than 100,000 deadweight tons but less than 500,000 dwt. Compare *ultralarge crude carrier*.

vessel experience factor *n:* a calculated factor, based on the experience of recent voyages, which reflects the average difference between vessel and shore measurements in terms of percent.

vibration dampener *n:* 1. a device, positioned in the drill stem between the bit and the drill collars, that absorbs impact loads and vibration from the up-and-down motion of the drill stem. Vibration dampeners are designed to transmit torque while absorbing reciprocative loads that decrease the efficiency of the drill bit. Also called a shock sub (trade name). 2. a device affixed to an engine crankshaft to minimize stresses that result from torsional vibration of the crankshaft.

viscometer *n:* a device used to determine the viscosity of a substance; also called a viscosimeter.

viscosimeter *n:* See *viscometer*.

viscosity *n:* a measure of the resistance of a liquid to flow. Resistance is brought about by the internal friction resulting from the combined effects of cohesion and adhesion. The viscosity of petroleum products is commonly expressed in terms of the time required for a specific volume of the liquid to flow through an orifice of a specific size.

viscosity index *n:* an index used to establish the tendency of an oil to thin out at increasing temperatures. Reference oils are a highly paraffinic Pennsylvania oil, rated 100, and a Gulf Coast naphthenic oil, rated 0.

VLCC *abbr:* very large crude carrier.

voids *n pl:* cavities in a rock that do not contain solid material but may contain fluids.

volatile *adj:* readily vaporized.

volatility *n:* the tendency of a liquid to assume the gaseous state.

volt *n:* the unit of electric potential, voltage, or electromotive force in the metric system. Its symbol is V.

voltage *n:* potential difference or electromotive force, measured in volts.

voltmeter *n:* an instrument used to measure, in volts, the difference of potential in an electrical circuit.

volume meter *n:* See *positive-displacement meter*.

volumeter *n:* See *positive-displacement meter*.

volumetric correction *n:* an equation that can be used to accurately calculate casing pressure according to the expansion of gas rising in the annulus when a well is shut in.

volumetric efficiency *n:* actual volume of fluid put out by a pump, divided by the volume displaced by a piston or pistons (or other device) in the pump. Volumetric efficiency is usually expressed as a percentage. For example, if the pump pistons displace 300 cubic inches, but the pump puts only

291 cubic inches per stroke, then the volumetric efficiency of the pump is 97 percent.

vortex finder *n:* the short pipe in a cone-shaped separator that extends down into the cone body from the top and that forces the whirling stream of material to start downward toward the small end of the cone body.

VRU *abbr:* vapor recovery unit.

vs *abbr:* versus.

vug *n:* a cavity in a rock.

vuggy formation *n:* also called vugular formation. See *vugular formation*.

vugular formation *n:* a rock formation that contains vugs; a cavernous formation. See *vug*.

vugular porosity *n:* a secondary rock porosity formed by the dissolving of the more soluble portions of a rock in waters containing carbonic or other acids.

W *sym:* watt.

wait-and-weight method *n:* a well-killing method in which the well is shut in and the mud weight is raised the amount required to kill the well. The heavy mud is then circulated into the well, while at the same time the kick fluids are circulated out. So called because one shuts the well in and waits for the mud to be weighted before circulation begins.

waiting on cement *adj:* pertaining to the time when drilling or completion operations are suspended so that the cement in a well can harden sufficiently.

walking beam *n:* the horizontal steel member of a beam pumping unit, having rocking or reciprocating motion.

wall cake *n:* also called filter cake or mud cake. See *mud cake*.

wall hook *n:* a device used in fishing for drill pipe. If the upper end of the lost pipe is leaning against the side of the wellbore, the wall hook centers it in the hole so that it may be recovered with an overshot, which is run on the fishing string and attached to the wall hook.

wall-hook guide *n:* See *wall hook*.

wall-hook packer *n:* also called a hook-wall packer. See *hook-wall packer*.

wall sticking *n:* also called differential-pressure sticking. See *differential-pressure sticking*.

wall-stuck pipe *n:* See *differential-pressure sticking*.

washout *n:* 1. excessive wellbore enlargement caused by solvent and erosional action of the drilling fluid. 2. a fluid-cut opening caused by fluid leakage.

washover *n:* the operation during which stuck drill stem or tubing is freed.

wash over *v:* to release pipe that is stuck in the hole by running washover pipe. The washover pipe must have an outside diameter small enough to fit into the borehole but an inside diameter large enough to fit over the outside diameter of the stuck pipe. A rotary shoe, which cuts away the formation, mud, or whatever is sticking the pipe, is made

Workover rig

up on the bottom joint of the washover pipe, and the assembly is lowered into the hole. Rotation of the assembly frees the stuck pipe. Several washovers may have to be made if the stuck portion is very long.

washover assembly n: See *washover pipe*.

washover back-off connector tool n: a fishing tool that is made up in a length of washover pipe connected to the top of the fish once the washover is completed, and then backed off the fish, thus enabling the washed-over portion of the fish to be retrieved. The tool permits washover, back-off, and pulling to be carried out in one round trip.

washover pipe n: an accessory used in fishing operations to go over the outside of tubing or drill pipe stuck in the hole because of cuttings, mud, and so forth that have collected in the annulus. The washover pipe cleans the annular space and permits recovery of the pipe. It is sometimes called washpipe.

washpipe n: 1. a short length of surface-hardened pipe that fits inside the swivel and serves as a conduit for drilling fluid through the swivel. 2. sometimes used to mean washover pipe. See *washover pipe*.

wash tank n: a tank containing heated water, through which crude-oil emulsion is forced to flow, and used to remove water from the crude. Also called a gun barrel.

water-back v: 1. to reduce the weight or density of a drilling mud by adding water. 2. to reduce the solids content of a mud by adding water.

water-base mud n: a drilling mud in which the continuous phase is water. In water-base muds, any additives are dispersed in the water. Compare *oil-base mud*.

water block n: a reduction in the permeability of a formation, caused by the invasion of water into the pores.

water-cement ratio n: the ratio of water to cement in a slurry. It is expressed as a percentage, indicating the number of pounds of water needed to mix 100 pounds of cement.

watercourse n: a hole inside a bit through which drilling fluid from the drill stem is directed.

water-cut paste n: a material that changes color (usually to red) in water. The use of water-cut paste is one method by which the level of water in the bottom of an oil storage tank can be determined. The paste is applied to a plumb bob, which is lowered to the bottom of the tank and then retrieved. The water level is then measured off the bob by noting the depth of the red portion of the bob.

water distillation unit n: equipment used mostly on offshore or desert locations to convert salt water to potable fresh water by distillation.

water drive n: the reservoir drive mechanism in which oil is produced by the expansion of the underlying water, which forces the oil into the wellbore. In general, there are two types of water drive: bottom-water drive, in which the oil is totally underlain by water, and edgewater drive, in which only the edge of the oil is in contact with the water. Bottom-water drive is more efficient.

watered-out adj: of a well, producing mostly water ("gone to water").

water encroachment n: the movement of water into a producing formation as the formation is depleted of oil and gas by production.

waterflood n: a method of secondary recovery in which water is injected into a reservoir to remove additional quantities of oil that have been left behind after primary recovery. Usually, waterflood involves the injection of water through wells specially set up for water injection and the removal of water and oil from the wells drilled adjacent to the injection wells.

waterflooding n: a method of enhanced oil recovery in which water is injected into a reservoir to remove additional quantities of oil that have been left behind after primary recovery. Usually, waterflooding involves the injection of water through wells specially set up for water injection and the removal of water and oil from production wells drilled adjacent to the injection wells.

water loss n: See *fluid loss*.

water maker n: See *water distillation unit*.

water string n: a string of casing used to shut off water above an oil sand.

water table n: 1. the structure at the top of the drilling derrick or mast that supports the crown block. 2. the underground level at which water is found.

watertight door n: a door on ships or mobile offshore rigs that, when closed, blocks the passage of water and withstands its pressure.

water-wet rock n: See *wettability*.

water zone n: the portion of an oil or gas reservoir occupied by water, usually the lowest zone in the reservoir.

watt *n:* the unit of power in the metric system. Its symbol is W.

wax *n:* See *paraffin*.

Wb *sym:* weber.

WC *abbr:* wildcat; used in drilling reports.

W/C *abbr:* water cushion; used in drilling reports.

wear sleeve *n:* a hollow cylindrical device attached to a joint of drill pipe midway between the tool joints that minimizes wear on the outside of the pipe at points where the pipe touches the wall of the hole.

weathering test *n:* a GPA test for LP gas for the determination of heavy components in a sample by evaporation of the sample as specified.

weber *n:* the metric unit of magnetic flux. Its symbol is Wb.

wedge factor *n:* a formula for trigonometric calculation to determine ROB when the vessel is out of trim and the ROB does not contact both forward and aft bulkheads. The formula assumes ROB to be a flowable liquid.

weevil *n:* shortened form of boll weevil. See *boll weevil*.

weight indicator *n:* an instrument near the driller's position on a drilling rig. It shows both the weight of the drill stem that is hanging from the hook (hook load) and the weight that is placed on the bit by the drill collars (weight on bit).

weighting material *n:* a material that has a high specific gravity and is used to increase the density of drilling fluids or cement slurries.

weight on bit *n:* the difference between the net weight of the entire drill stem and the reduced weight resulting when the bit is resting on bottom.

weight up *v:* to increase the weight or density of drilling fluid by adding weighting material.

weir *n:* a device installed in a separator, treater, or mud tank and used to determine the amount of liquid flowing over it when the depth of the liquid is known.

well *n:* the hole made by the drilling bit, which can be open, cased, or both. Also called wellbore, borehole, or hole.

wellbore *n:* a borehole; the hole drilled by the bit. A wellbore may have casing in it or it may be open (uncased); or a portion of it may be cased, and a portion of it may be open. Also called borehole or hole.

wellbore soak *n:* an acidizing treatment in which the acid is placed in the wellbore and allowed to react by merely soaking; also called wellbore cleanup. It is a relatively slow process because very little of the acid actually comes in contact with the formation.

well completion *n:* the activities and methods necessary to prepare a well for the production of oil and gas; the method by which a flow line for hydrocarbons is established between the reservoir and the surface. The method of well completion used by the operator depends on the individual characteristics of the producing formation or formations. Such techniques include open-hole completions, sand-exclusion completions, tubingless completions, multiple completions, and miniaturized completions.

well control *n:* the methods used to prevent a well from blowing out. Such techniques include, but are not limited to, keeping the borehole completely filled with drilling mud of the proper weight or density during all operations, exercising reasonable care when tripping pipe out of the hole to prevent swabbing, and keeping careful track of the amount of mud put into the hole to replace the volume of pipe removed from the hole during a trip.

wellhead *n:* the equipment installed at the surface of the wellbore. A wellhead includes such equipment as the casinghead and tubing head. *adj:* pertaining to the wellhead (e.g., wellhead pressure).

well logging *n:* the recording of information about subsurface geologic formations. Logging includes records kept by the driller and records of mud and cutting analyses, core analysis, drill stem tests, and electric, acoustic, and radioactivity procedures. See *driller's log, mud analysis, core analysis, drill stem test, electric well log, acoustic log,* and *radioactivity log*.

well permit *n:* authorization, usually by a governmental conservation agency, to drill a well. A permit is sometimes required also for deepening or remedial work.

well puller *n:* a member of a well-servicing crew.

well-servicing *adj:* relating to well-servicing work, (e.g., a well-servicing company).

well servicing *n:* the maintenance work performed on an oil or gas well to improve or maintain the production from a formation already producing. Usually, it involves repairs to the pump, rods, gas-lift valves, tubing, packers, and so forth.

well site *n:* also called location. See *location*.

well spacing *n:* regulation for conservation purposes of the number and location of wells over a reservoir.

well stimulation *n:* any of several operations used to increase the production of a well. See *acidize* and *formation fracturing*.

well surveying *n:* also called well logging. See *well logging*.

wet box *n:* also called a mud box. See *mud box*.

wet gas *n:* 1. a gas containing water, or a gas that has not been dehydrated. 2. also called rich gas. See *rich gas*.

wet oil *n:* an oil that contains water, either as an emulsion or as free water.

wet suit *n:* a diving suit, usually made of neoprene material, designed to provide thermal insulation for a diver's body. A small amount of water that enters the suit is warmed by body heat and provides protection for dives of short duration.

wettability *n:* the relative affinity between individual grains of rock and each fluid that is present in the spaces between the grains. If oil and water are both present, the water is usually in contact with the surface of each grain, and the rock is called water-wet. However, if the oil contacts the surface, the rock is called oil-wet.

wetting *n:* the adhesion of a liquid to the surface of a solid.

wet welding *n:* underwater welding performed without the use of a protective habitat.

wh *abbr:* white; used in drilling reports.

wheel-type back-off wrench *n:* a wheel-shaped wrench that is attached to the sucker rod string at the surface and is manually turned to unscrew the string to allow it to be pulled from the well. Also called a back-off wheel or a circle wrench.

whelp *n:* (nautical) a sprocket tooth in a wildcat.

whipstock *n:* a long steel casing that uses an inclined plane to cause the bit to deflect from the original borehole at a slight angle. Whipstocks are sometimes used in controlled directional drilling, in straightening crooked boreholes, and in sidetracking to avoid unretrieved fish.

wickers *n:* broken or frayed strands of the steel wire that comprises the outer wrapping of wire rope.

wildcat *n:* 1. a well drilled in an area where no oil or gas production exists. With present-day exploration methods and equipment, about one wildcat out of every nine proves to be productive, although not necessarily profitable. 2. (nautical) the geared sheave of a windlass used to pull anchor chain. *v:* to drill wildcat wells.

wildcatter *n:* one who drills wildcat wells.

wild well *n:* a well that has blown out of control and from which oil, water, or gas is escaping with great force to the surface; also called a gusher.

winch *n:* a machine that pulls or hoists by winding a cable around a spool.

windbreak *n:* something that breaks the force of the wind. For example, canvas windbreaks installed around the outside of the rig floor on a drilling or workover rig afford the crew protection from strong, cold winds. Sometimes called a prefab.

wind girder *n:* See *wind ring*.

wind guy line *n:* the wire rope attached to ground anchors to provide lateral support for a mast or derrick. Compare *load guy line*.

windlass *n:* a device on an anchor-handling boat that propels the anchor chain to and from a chain locker where it is stored.

wind-load rating *n:* a specification used to indicate the resistance of a derrick to the force of wind. The wind-load rating is calculated according to API specifications. Typical wind resistance in derricks is 75 mph with pipe standing in the derrick and 115 mph or higher without.

window *n:* a slotted opening or a full section removed in the pipe lining (casing) of a well, usually made to permit sidetracking.

wind ring *n:* a horizontal stiffening and structural member installed near the top of a floating-roof tank to reinforce the tank wall against wind pressure; also called a wind girder.

windward *n:* (nautical) upwind; the direction from which the wind is blowing.

wiper *n:* a circular rubber device with a split in its side that is put around drill pipe and serves to wipe or clean drilling mud off the outside of the pipe as the pipe is pulled from the hole.

wireline *n:* a small-diameter metal line used in wireline operations; also called slick line. Compare *conductor line*.

wireline core barrel *n:* See *core barrel*.

wireline cutting tool *n:* a device usually run on a solid wireline, used to cut another wireline stuck in a well.

wireline formation tester *n:* a formation fluid sampling device that also logs flow and shut-in pressure in rock near the borehole. A spring mechanism holds a pad firmly against the sidewall while a piston creates a vacuum in a test chamber. Formation fluids enter the test chamber through a valve in the pad. A recorder logs the rate at which the test chamber is filled. Fluids may also be drawn to fill a sampling chamber. Wireline formation tests may be done any number of times during one trip in the hole, so they are very useful in formation testing.

wireline operations *n pl:* the lowering of mechanical tools, such as valves and fishing tools, into the well for various purposes. Electric wireline operations, such as electric well logging and perforating, involve the use of conductor line, which in the oil patch is commonly but erroneously called wireline.

wireline probe *n:* a diagnostic tool used to ascertain the position of a gas leak in the tubing of a gas-lift well.

wireline spear *n:* a special fishing tool fitted with prongs to catch and recover wireline that has broken off and been left in a well.

wireline survey *n:* a general term often used to refer to any type of log being run in a well.

wireline well logging *n:* the recording of subsurface characteristics by wireline tools. Wireline well logs include acoustic logs, caliper logs, radioactivity logs, and resistivity logs. See *acoustic well logging, caliper log, radioactivity well logging,* and *resistivity well logging.*

wireline wiper *n:* a flexible rubber wiper used to scrape mud or oil from a wireline as it is pulled from a hole.

wire rope *n:* a cable composed of steel wires twisted around a central core of fiber or steel wire to create a rope of great strength and considerable flexibility. Wire rope is used as drilling line (in rotary and cable-tool rigs), coring line, servicing line, winch line, and so on. It is often called cable or wireline; however, wireline is a single, slender metal rod, usually very flexible. Compare *wireline.*

wobble *n:* movement between the mating surfaces of box and pin in a tool joint. *v:* to move in a rocking motion.

WOC *abbr:* waiting on cement; used in drilling reports.

WO/O *abbr:* waiting on orders; used in drilling reports.

WOR *abbr:* water-oil ratio.

working barrel *n:* the outer shell of a downhole plunger pump. The pumping cycle starts with an upward stroke of the rods, which pulls the plunger up through the working barrel. On the upstroke, the traveling valve closes, the standing valve in the working barrel opens, fluid above the traveling valve is lifted out of the well, and a new charge is drawn into the pump. On the downstroke, the traveling valve opens, the standing valve closes, and the fluid is forced from the working barrel through the traveling valve in the plunger and into the tubing. Repeated strokes bring the fluid to the surface.

working interest *n:* the portion of oil production money out of which operating and development costs are paid (i.e., the portion remaining after deduction of royalty interest).

working-interest oil *n:* See *net production* and *working interest.*

working pressure *n:* the maximum pressure at which an item is to be used at a specified temperature.

workover *n:* the performance of one or more of a variety of remedial operations on a producing oilwell to try to increase production. Examples of workover jobs are deepening, plugging back, pulling and resetting liners, squeeze cementing, and so forth.

work over *v:* to perform one or more of a variety of remedial operations on a producing oilwell to try to increase production. Examples of workover operations are deepening, plugging back, pulling and resetting liners, squeeze cementing, and so on.

workover fluid *n:* a special drilling mud used to keep a well under control while it is being worked over. A workover fluid is compounded carefully so that it will not cause formation damage.

workover rig *n:* a portable rig used for working over a well. See *production rig.*

work string *n:* the string of drill pipe or tubing suspended in a well to which is attached a special tool or device that is used to carry out a certain task, such as squeeze cementing or fishing.

worm *n:* a new and inexperienced worker on a drilling rig.

worm gear *n:* the gear of a worm (a short revolving screw with spiral-shaped threads) and a worm wheel (a toothed wheel gearing the thread of a worm) working together.

wrench flat *n:* a flat area on a otherwise round fitting, to which a wrench can be applied (as on sucker rod coupling). Also called a wrench square.

wrench square *n:* See *wrench flat.*

wrist pin *n:* See *piston pin.*

XYZ

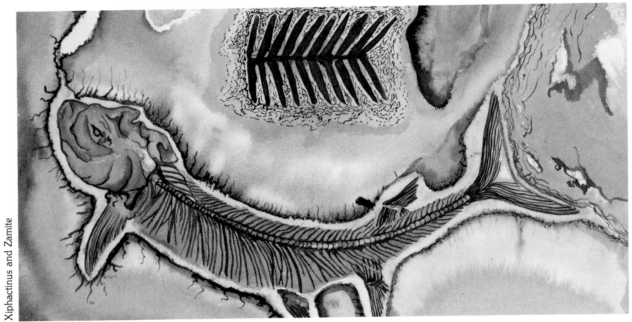

Xiphactinus and Zamite

xln *abbr:* crystalline; used in drilling reports.

xylene *n:* any of three flammable hydrocarbons, $C_6H_4(CH_3)_2$, similar to benzene. A commercial mixture is used as a solvent when oil field emulsions are being tested.

yaw *n:* on a mobile offshore drilling rig or ship, the angular motion about a line perpendicular to a horizontal plane through the rig or ship; the angular motion as the bow or stern moves from side to side. *v:* to move from side to side (as a ship).

yd *abbr:* yard.

yd² *abbr:* square yard.

yd³ *abbr:* cubic yard.

yield point *n:* the maximum stress that a solid can withstand without undergoing permanent deformation either by plastic flow or by rupture. See *tensile strength*.

yield strength *n:* a measure of the force needed to deform tubular goods to the extent that they are permanently distorted.

zone *n:* a rock stratum that is different from or distinguished from another stratum (e.g., a pay zone).

zone of lost circulation *n:* a formation that contains holes or cracks large enough to allow cement to flow into the formation instead of up along the annulus outside of the casing.

ABBREVIATIONS

A: ampere
AC: alternating current
acre: (not abbreviated)
A-h: ampere-hour
atm: atmosphere
avg: average

bbl: barrel
Bcf: billion cubic feet
Bcf/d, Bcf/D: billion cubic feet per day
bbl/d, B/D, b/d, BPD, bpd: barrels per day
bhp: brake horsepower
BHP: bottomhole pressure
BHT: bottomhole temperature
BLPD: barrels of liquid per day
BOPD: barrels of oil per day
Bscf/d, Bscf/D: billion standard cubic feet per day
Btu: British thermal unit
BWPD: barrels of water per day

°C: degrees Celsius
cal: calorie
cm: centimetre
cm^2, sq cm: square centimetre
cm^3, cc: cubic centimetre
cp: centipoise

d, D: day
D, darcy: darcy
DC: direct current
°API: degrees API (American Petroleum Institute)
°F: degrees Fahrenheit
dm: decimetre
dm^3: cubic decimetre
dm^3/s: cubic decimetres per second

emf: electromotive force

fathom: (not abbreviated)
ft: foot
ft-lb: foot-pound
ft/min, fpm: feet per minute
ft/s, fps: feet per second
ft^2, sq ft: square foot
ft^3, cu ft: cubic foot
ft^3/bbl, cu ft/bbl: cubic feet per barrel
ft^3/d, cu ft/d, cfd, cfD: cubic feet per day
ft^3/lb, cu ft/lb, cfp: cubic feet per pound
ft^3/min, cu ft/min, cfm: cubic feet per minute
ft^3/s, cu ft/s, cfs: cubic feet per second

g: gram
gal: gallon
gpm, gal/min: gallons per minute
GOR: gas-oil ratio

h: hour
hp: horsepower
hp-h, hp-hr: horsepower-hour
Hz: hertz

ID: inside diameter
in.: inch
in.^{2}, sq in.: square inch
in.^{3}, cu in.: cubic inch
in./s, ips: inches per second

J: joule

K: kelvin
kg: kilogram
km: kilometre
kPa: kilopascal
kV, kv: kilovolt
kW, kw: kilowatt
kW-h, kwh: kilowatt-hour

lb: pound
lb/ft^3, pcf: pounds per cubic foot
L: litre
LPG: liquefied petroleum gas

m: metre
m^2, sq m: square metre
m^3, cu m: cubic metre
m^3/d: cubic metres per day
mA, milliamp: milliampere
Mcf: thousand cubic feet
Mcf/d, Mcf/D: thousand cubic feet per day
md: millidarcy
mev: million electron volts
MER: maximum efficiency rate
mg: milligram
μs, microsec: microsecond
mile: (not abbreviated)
min: minute
ml: millilitre
mm: millimetre
MMcf: million cubic feet
MMcf/d, MMcf/D: million cubic feet per day
MMscf/d, MMscf/D: million standard cubic feet per day

mol: mole
MPa: megapascal
mph: miles per hour
Mscf/d, Mscf/D: thousand standard cubic feet per day
mV, mv: millivolt

N: newton

Pa: pascal

OD: outside diameter
oz: ounce

PI: productivity index
ppg, lb/gal: pounds per gallon
ppm: parts per million
psi: pounds per square inch
psia: pounds per square inch absolute
psig: pounds per square inch gauge
PVT: pressure-volume-temperature

°R: degrees Rankine

s, sec: second
scf: standard cubic feet
scf/d, scf/D: standard cubic feet per day
S/m: siemens per metre
SP: spontaneous potential
spm: strokes per minute
SSU: Saybolt seconds universal
STB: stock-tank barrel
STB/d, STB/D: stock-tank barrels per day
std: standard

t: tonne
T: ton
Tcf: trillion cubic feet
Tcf/d, Tcf/D: trillion cubic feet per day

V, v: volt

W, w: watt
WOR: water-oil ratio

yd: yard
yd^2: square yard
yd^3: cubic yard

SI UNITS

Quantity	Unit Name	Symbol	Formula
Base Units			
Length	metre	m	
Mass	kilogram	kg	
Time	second	s	
Electric current	ampere	A	
Temperature	kelvin	K	
Amount of substance	mole	mol	
Luminous intensity	candela	cd	
Supplementary Units			
Plane angle	radian	rad	
Solid angle	steradian	sr	
Derived Units			
Area	square metre	m^2	
Volume	cubic metre	m^3	
Speed, velocity	metre per second	m/s	
Acceleration	metre per second squared	m/s^2	
Density	kilogram per cubic metre	kg/m^3	
Concentration	mole per cubic metre	mol/m^3	
Specific volume	cubic metre per kilogram	m^3/kg	
Luminance	candela per square metre	cd/m^2	
Moment of force	newton metre	N•m	
Derived Units with Special Names			
Frequency	hertz	Hz	1/s
Force	newton	N	$kg•m/s^2$
Pressure, stress	pascal	Pa	N/m^2
Energy, work, quantity of heat	joule	J	N•m
Power	watt	W	J/s
Electric charge	coulomb	C	A•s
Electric potential	volt	V	W/A
Electric resistance	ohm	Ω	V/A
Electric conductance	siemens	S	A/V
Electric capacitance	farad	F	C/V

Quantity	Unit Name	Symbol	Formula
Magnetic flux	weber	Wb	V·s
Inductance	henry	H	Wb/A
Magnetic flux density	tesla	T	Wb/m²
Luminous flux	lumen	lm	cd·sr
Illuminance	lux	lx	lm/m²
Activity of radionuclides	becquerel	Bq	s^{-1}
Absorbed dose of ionizing radiation	gray	Gy	J/kg
Non-SI Units Allowable with SI			
Time	minute	min	1 min = 60s
	hour	h	1 h = 3 600s
	day	d	1 d = 86 400s
	year	a	
Plane angle	degree	°	1° = π/180 rad
	minute	'	1' = π/10 800 rad
	second	"	1" = π/648 000 rad
Capacity or volume	litre	L	1 L = 1 dm³
Temperature	degree Celsius	°C	interval of 1°C = 1K
Mass	tonne	t	1 t = 1 000 kg
Revolution	revolution	r	1 r = 2 rad
Marine and aerial distance	nautical mile		1 nautical mile = 1 852 m
Marine and aerial velocity	knot	kn	1 nautical mile per hour = (1 852/3 600)m/s
Land area	hectare	ha	1 ha = 10 000 m²
Pressure	standard atmosphere	atm	1 atm = 101.325 kPa

SI UNITS FOR DRILLING

Quantity or Property	Conventional Units	SI Unit	Symbol	Multiply by
Depth	feet	metres	m	0.3048
Hole and pipe diameters Bit size	inches	millimetres	mm	25.4
Weight on bit	pounds	decanewtons	daN	0.445
Nozzle size	32nds inch	millimetres	mm	0.794
Drill rate	feet/hour	metres/hour	m/h	0.3048
Volume	barrels	cubic metres	m^3	0.1590
	US gals/stroke	cubic metres per stroke	m^3/stroke	3.785 L/stroke
Pump output and flow rate	US gpm	cubic metres per minute	m^3/min	0.00378
	bbl/stroke	cubic metres per stroke	m^3/stroke	An oil barrel is 0.1589873 m^3.
	bbl/min	cubic metres per minute	m^3/min	0.1590
Annular velocity Slip velocity	feet/min	metres per minute	m/min	0.3048
Liner length and diameter	inches	millimetres	mm	25.4
Pressure	psi	kilopascals	kPa	6.895
		megapascals	MPa	0.006895
Bentonite yield	bbl/ton	cubic metres per tonne	m^3/t	0.175
Particle size	microns	micrometres	μm	1
Temperature	°Fahrenheit	°Celsius	°C	(°F − 32)/1.8
Mud density	ppg (US)	kilograms per cubic metre	kg/m^3	119.82
Mud gradient	psi/foot	kilopascals per metre	kPa/m	22.621
Funnel viscosity	s/quart (US)	seconds per litre	s/L	1.057
Apparent and plastic viscosity	centipoise	millipascal seconds	mPa·s	1
Yield point Gel strength and stress	lb_f/100 ft^2	pascals	Pa	0.4788 (0.5 for field use)
Cake thickness	32nds inch	millimetres	mm	0.794
Filter loss	millimetres or cubic centimetres	cubic centimetres	cm^3	1

Quantity or Property	Conventional Units	SI Unit	Symbol	Multiply by
MBT (bentonite equivalent)	lb/bbl	kilograms per cubic metre	kg/m^3	2.85
Material concentration	lb/bbl	kilograms per cubic metre	kg/m^3	2.85
Shear rate	reciprocal seconds	reciprocal seconds	s^1	1
Torque	foot-pounds	newton metres	N·m	1.3358
Table speed	revolutions per minute	revolutions per minute	r/min	1
Ionic concentration in water	equivalents per million	moles per cubic metre	mol/m^3	1
Corrosion rates	lb/ft^2/year	grams per square metre per day	g/m^2·d	13.377
	mils per year	millimetres per year	mm/a	0.0254

METRIC EQUIVALENTS

Length

1 millimetre = 0.04 inch
1 centimetre = 0.39 inch
1 metre = 39.37 inches = 1.09 yards

1 inch = 2.54 centimetres
1 foot = 3.05 decimetres
1 yard = 0.91 metre
1 mile = 1.61 kilometres

Area

1 square centimetre = 0.15 square inch
1 square decimetre = 0.11 square foot
1 square metre = 1.20 square yards
1 hectare = 2.47 acres
1 square kilometre = 0.39 square mile

1 square inch = 6.45 square centimetres
1 square foot = 9.29 square decimetres
1 square yard = 0.83 square metre
1 acre = 0.40 hectare
1 square mile = 2.59 square kilometres

Pressure

1 kilopascal = 0.145 pound per square inch
1 kilopascal per metre = 0.044 pound per square inch per foot
1 pound per square inch = 6.894 kilopascals
1 pound per square inch per foot = 22.62 kilopascals per metre

Volume

1 cubic centimetre = 0.06 cubic inch
1 cubic metre (stere) = 1.31 cubic yards

1 cubic inch = 16.39 cubic centimetres
1 cubic foot = 0.28 cubic decimetre
1 cubic yard = 0.75 cubic metre

Capacity

1 millilitre = 0.06 cubic inch
1 litre = 61.02 cubic inches = 1.507 liquid quarts
1 decalitre = 0.35 cubic foot = 2.64 liquid gallons

1 fluid ounce = 29.57 millilitres
1 U.S. gallon = 3.785 litres
1 barrel (oil) = 159 litres

Weight

1 gram = 0.04 ounce
1 kilogram = 2.20 pounds
1 metric ton (tonne) = 0.98 English ton

1 ounce = 28.35 grams
1 pound = 0.45 kilogram
1 English ton = 1.02 metric tons

Density

1 kilogram per litre = 8.34 pounds per gallon
1 kilogram per litre = 62.5 pounds per cubic foot

1 pound per gallon = 0.119 kilogram per litre
1 pound per cubic foot = 0.016 kilogram per litre

METRIC PREFIXES

Value	Prefix	Symbol
10^{18}	exa	E
10^{15}	peta	P
10^{12}	tera	T
10^{9}	giga	G
10^{6}	mega	M
10^{3}	kilo	k
10^{2}	hecto	h
10^{1}	deca	da

Value	Prefix	Symbol
10^{-1}	deci	d
10^{-2}	centi	c
10^{-3}	milli	m
10^{-6}	micro	μ
10^{-9}	nano	n
10^{-12}	pico	p
10^{-15}	femto	f
10^{-18}	atto	a